THE RAVEN ESSAYS

Indigenous Environmental Justice, Education, and Self-Determination

Edited by John Borrows, Dawn Hoogeveen, Max Ritts, and Susan Smitten

Named after the Respecting Aboriginal Values and Environmental Needs (RAVEN) nonprofit organization, *The RAVEN Essays* is an anthology that celebrates a decade of prize-winning student essays. Since 2012, RAVEN has awarded an annual essay prize to honour students who champion the vital importance of Indigenous rights and self-determination, both in Canada and globally. The essays featured in this collection highlight exceptional student work while reflecting on the evolving relationship between Indigenous politics and academia. From issues like fishing rights and the Trans Mountain Pipeline to challenges of sexism and conservation policy, these essays capture a transformative period in Indigenous struggles, offering insights that resonate far beyond the Canadian settler state.

The anthology also includes contributions from prominent scholars such as Glen Coulthard, Dara Culhane, Michael Fabris, Sarah Hunt/Tłaliłila'ogwa, and Heather Dorries. Five complementary essays explore various aspects of structural change, institutional constraints, and broader commitments to Indigenous knowledge within university settings. Aimed at readers in Indigenous law, environmental studies, anthropology, and geography, *The RAVEN Essays* is a book created by students for students, and by academics for the academy.

Together, the contributors reflect on the powerful formation and enactment of Indigenous law, environmental stewardship, place-based knowledge, pedagogy, and literacy – both within the academy and in the broader community, across land, water, and culture.

JOHN BORROWS is a professor and the Loveland Chair in Indigenous Law in the Faculty of Law at the University of Toronto.

DAWN HOOGEVEEN is a research associate in the Faculty of Health Sciences at Simon Fraser University.

MAX RITTS is an assistant professor in the Graduate School of Geography at Clark University.

SUSAN SMITTEN is an award-winning filmmaker and writer; she is retired from her role as the executive director of RAVEN (Respecting Aboriginal Values and Environmental Needs).

The RAVEN Essays

Indigenous Environmental Justice, Education, and Self-Determination

EDITED BY JOHN BORROWS,
DAWN HOOGEVEEN, MAX RITTS,
AND SUSAN SMITTEN

UNIVERSITY OF TORONTO PRESS
Toronto Buffalo London

© University of Toronto Press 2025
Toronto Buffalo London
utorontopress.com
Printed in Canada

ISBN 978-1-4875-6237-3 (cloth)   ISBN 978-1-4875-6240-3 (EPUB)
ISBN 978-1-4875-6238-0 (paper)   ISBN 978-1-4875-6239-7 (PDF)

Library and Archives Canada Cataloguing in Publication

Title: The RAVEN essays : Indigenous environmental justice, education, and self-
    determination / edited by John Borrows, Dawn Hoogeveen, Max Ritts, and Susan
    Smitten.
Other titles: Indigenous environmental justice, education, and self-determination
Names: Borrows, John, 1963– editor | Hoogeveen, Dawn, editor. |
    Ritts, Max, 1982– editor. | Smitten, Susan, 1961– editor
Description: Includes bibliographical references and index.
Identifiers: Canadiana (print) 20240539818 | Canadiana (ebook) 2024054000X | ISBN
    9781487562373 (hardcover) | ISBN 9781487562380 (softcover) | ISBN 9781487562403
    (EPUB) | ISBN 9781487562397 (PDF)
Subjects: LCSH: Environmental justice – Canada. | LCSH: Environmental protection –
    Canada. | LCSH: Environmental policy – Social aspects – Canada. | LCSH: Indigenous
    peoples – Land tenure – Canada. | LCSH: Traditional ecological knowledge – Canada. |
    LCSH: Indigenous peoples – Legal status, laws, etc. – Canada.
Classification: LCC GE240.C3 R38 2025 | DDC 333.7089/97071 – dc23

Cover design: Mark Rutledge
Cover image: Adrian Nadjiwon, *Ravens and Moon*, 2024, digital illustration

We wish to acknowledge the land on which the University of Toronto Press operates. This
land is the traditional territory of the Wendat, the Anishnaabeg, the Haudenosaunee, the
Métis, and the Mississaugas of the Credit First Nation.

This book has been published with the help of a grant from the Federation for the
Humanities and Social Sciences, through the Awards to Scholarly Publications Program,
using funds provided by the Social Sciences and Humanities Research Council of Canada.

University of Toronto Press acknowledges the financial support of the Government of
Canada, the Canada Council for the Arts, and the Ontario Arts Council, an agency of the
Government of Ontario, for its publishing activities.

Canada Council   Conseil des Arts
for the Arts   du Canada

ONTARIO ARTS COUNCIL
CONSEIL DES ARTS DE L'ONTARIO
an Ontario government agency
un organisme du gouvernement de l'Ontario

Funded by the   Financé par le
Government   gouvernement
of Canada   du Canada

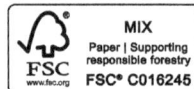

MIX
Paper | Supporting
responsible forestry
FSC® C016245

About the Cover Artist

Adrian is a student both of the visual arts and of his own Native heritage. This curiosity was piqued at a young age by images painted in the Woodland style, a manner of painting developed by Cree and Anishinaabe artists such as Norval Morrisseau, Ahmoo Angeconeb, and Roy Thomas. Inspired by those colourful images, Adrian began a journey to understand the roots of this art, his ancestry, and spirituality – a journey that continues to this day.

Contents

Part Two: Relations

Part Three: Struggles

List of Illustrations

Acknowledgments

The authors would like to respectfully acknowledge that much of this work took place on the territories of the Esquimalt (Xwsepsum) and Songhees Nations, the xwməθkwəy̓əm (Musqueam), S̱ḵwx̱wú7mesh (Squamish), and Selíĺwitulh (Tsleil-Waututh) Nations. This collection would not have been possible without the students who contributed their essays, nor without the creativity of RAVEN as an organization and its efforts to make space for student engagement through the essay prize. We are very grateful for the time and energy of the students and the passion and expertise they shared with us through their work. Along with the judges who participated in the roundtable (Mike Fabris, Glen Coulthard, and John Borrows), we would also like to acknowledge the long-standing contributions of Dara Culhane, who volunteered her time and served as an adjudicator for the student essay prize across several of the award's iterations.

THE RAVEN ESSAYS

The RAVEN Essays

SUSAN SMITTEN

As with so many great initiatives, it starts with a twinkle in one person's eye, the inspiration that dances them into the creative space. In this specific instance, the eye belonged to a young geography student who presented the idea of an essay prize for undergraduate scholars in 2011 to a fledgling charity still growing its pinions. RAVEN was barely two years old but already living up to its name. Many Pacific Northwest tribes view Raven as a trickster who takes on jobs no one else wants to do, and by doing so reshapes and transforms the world. Max Ritts saw that students working on academic papers were generally not rewarded within the academy for the out-of-the-box and often controversial thinking so required to move the needle on Indigenous rights and on issues of reconciliation. He suggested that RAVEN might host an award to honour critical thinking and academic achievement in areas that have been overlooked.

A year before this, I co-founded an organization with an institutional goal of providing legal aid to Indigenous peoples struggling to assert their sovereignty in the lands and waters known today as Canada. When I met Max, RAVEN, or Respecting Aboriginal Values and Environmental Needs, did not yet have a firm grasp on its niche mandate of fundraising to assist Indigenous nations with access to justice, so taking on this random yet intriguing idea was first more problem than promise. But ravens are acrobatic fliers. They have a reputation for solving ever more complex problems with creativity and ingenuity. In the vacuum of resources to support the idea, Max brought in Dawn Hoogeveen, and the pair agreed to convene the adjudicating panel and send out posters to universities, while the tiny team of two at RAVEN hustled to find the prize funds. RAVEN chose to support the prize because it resonated deeply with the organization's core values of learning – being curious and creating opportunities to share knowledge and expand understanding. It also aligned with our principle of knowledge and trust – building a cycle of learning that expands understanding of Indigenous rights and title and sharing diverse opinions and views with dignity and respect. To an important degree, the subjects broached in this collection speak to the wide areas of concern that define the RAVEN legal agenda: protecting treaty rights, stopping unjust fossil fuel expansion, protecting nation-specific relations to animal kin, enshrining rights for future Indigenous generations.

The first winner, selected in 2012, was Johnnie Manson with his essay "The Dialectics of First Nation Governance: A Case Study of the Relationship between the Ahousaht First Nation and Industrial Fish Farming." That started an annual adventure into academia with topics ranging from decolonization and contemporary colonization to a personal look at Bill C-3 and Indigenous enfranchisement and Indigenous teachings about food practices in a time of climate crisis.

And as the essay prize evolved, RAVEN grew into its legal theory of change, fundraising for legal actions led by more than a dozen Indigenous nations. These campaigns resulted in critical achievements for Indigenous rights and the environment: cancellation of the Enbridge Northern Gateway Pipeline; protection of 83 per cent of the Peel watershed in the Yukon; defence of Teztan Biny (Fish Lake) and T'ak Tl'ah Bin (Morrison Lake) from proposed mining developments; cancellation of the Petronas Pacific Northwest LNG project at the mouth of the Skeena River; an evolving effort to hold government and corporate powers responsible for an oil spill on Heiltsuk territory; nationwide support for Beaver Lake Cree Nation's continued pursuit of environmental justice to stop degradation of traditional lands by tar sands industries; and opposition by Wet'suwet'en hereditary chiefs to a proposed LNG pipeline.

So much of what RAVEN's partner nations are trying to achieve is the laying aside of unjust, outdated colonial policies – the reining in of Crown sovereignty from within the system. Their empowerment is as much about challenging deeply flawed policies that further subjugate and dispossess Indigenous peoples today as it is about a legal winning streak. Nevertheless, RAVEN's support means that Indigenous nations do not have to divert funds away from critical community resources like housing, education, resource management, infrastructure, and mental health care to pursue social justice through the courts.

It is worth noting that at the time of writing, more than twelve years after RAVEN began operations, it remains the only non-profit charity in Canada with a mandate to raise legal defence funds to assist Indigenous peoples who choose to access the court system in order to protect their rights, lands, waters, and way of life. RAVEN is also mandated to assist

Indigenous peoples in protecting and restoring the natural environment for the benefit of all people within what is now called Canada by developing and delivering education programs to advance knowledge and understanding of available legal rights and remedies.

We are the legal agents in this world. We all have an opportunity to practise law. Sometimes that is done by standing side by side with Indigenous peoples, and in some instances, providing resources to help Indigenous peoples as they are raising their voice. So, providing resources is actually a practice of law – it's a custom that can be used to put us in relationship with one another.

– John Borrows

In many Indigenous communities that have been deeply affected by extractive industries, there exists a significant gap in resources to advocate with the legal rights available that can effectively address cultural and ecological issues. A well-resourced community that acknowledges the relationship between cultural integrity and a healthy ecology is able to act clearly and decisively to ensure its long-term viability. RAVEN now works together with the Harmony Foundation of Canada to present the Environmental Essay Prize for Young Scholars annually in support of students advocating for Indigenous rights and self-determination.

The purpose of the prize is to recognize outstanding work that explores cutting-edge thinking while presenting cases where Indigenous knowledge inspires practical action and produces meaningful results with social, cultural, environmental, and/or economic benefits. RAVEN works to recognize students who are seeking constructive responses to important topics, such as the factors shaping Indigenous youth movements today, the social geography giving rise to Indigenous action, the struggle to save sacred headwaters, attempts to incorporate Indigenous traditional ecological knowledge into environmental assessment processes, lands and rights, and the role of colonialism and the *Indian Act* in producing systemic inequity.

Among RAVEN's core values are humility, determination, and reconciliation, and as evidenced by the papers in this collection, approaches to stewardship that take into account the wisdom of the past, current scientific and Indigenous knowledge, and a vision for future generations are a powerful way to secure our common future. We are deeply grateful to Max and Dawn for their commitment to and passion for the essay prize and their work toward the creation of this book, and to John Borrows for his enthusiasm and many contributions to making it a reality. Also, deep gratitude to the authors, adjudicators, editors, and the team behind the scenes making this collection real and readable. Finally, to readers: Enjoy!

Chapter One

Situating the RAVEN Essays

DAWN HOOGEVEEN, MAX RITTS, AND HEATHER DORRIES

This collaboration flows from a decade of adjudicating the RAVEN scholar award. In this chapter, we consider the contribution of student activism and scholarship to social and environmental justice. As scholars who have worked in Indigenous studies and geography, we consider the significance of student activism and academic work that guide essays within select academic fields. While RAVEN assists Indigenous peoples in pursuing legal challenges across Canada, this work is also relevant to the resurgence of Indigenous laws more broadly. Furthermore, while the essays gathered here highlight scholarly debates surrounding Indigenous politics in a settler-colonial state, they also show how broader political, social, and environmental forces are reflected in critical scholarship.

This anthology points to a promising development in contemporary academic life: the surge of student-led activism around Indigenous rights to self-determination and environmental justice, decolonization, and sovereignty. This movement stretches well beyond the last decade of university-based activism in Canada and shows no signs of letting up. Movements that prioritize Indigenous self-determination and decolonization are significant, both inside and beyond the university. The university has been a key site of the production and circulation of racist forms of knowledge embedded in colonization and processes of dispossession. As Linda Tuhiwai Smith astutely noted in her path-breaking work *Decolonizing Methodologies: Research and Indigenous Peoples,* "imperialism has been perpetuated through the ways in which knowledge about indigenous peoples was collected, classified and then represented in various ways back to the West."[1] Imperial classification, as noted by Smith, is central within academic institutions and traditional research methods. Smith's work lays out an agenda for the decolonization of research that relies not only on the proliferation of Indigenous theory, but also Indigenous-led, community-based action. This collection demonstrates ways in which Indigenous studies works to counter imperialist colonial norms. Indigenous sovereignty movements represent not only a political orientation, but also an "epistemological challenge" to the ways that that "truths" about Indigenous people have been constructed and circulated in the academy.[2] Movements to reclaim Indigenous political authority are locally articulated, but also concern the broad reclamation of "intellectual sovereignty"[3] that moves against the ways knowledge about Indigenous peoples has been produced and circulated across academic contexts.

As precursor to the current moment and last decade of student activism displayed here, Red Power activisms transformed universities on both sides of the border in the 1970s – from Trent University in Williams Treaty territory to Princeton in the United States.[4] It carries the "returns" of more recent political forces too, such as the Indigenous-led movements of Idle No More (2012–13) and the Dakota Access Pipeline protests (2017), both of which are directly addressed in subsequent chapters. Indigenous-led student activism is animated by campus activities. Indigenous students and student allies are joining Indigenous-led social and environmental justice movements: from and anti-pipeline struggles, to healing gardens, language revitalization, and increasing questions of how best to acknowledge the Indigenous territories upon which Canadian institutions sit.[5] The efforts of Indigenous scholars and student activists are slowly transforming the university and are, to some extent, resonating within the walls of these institutions. Today, many universities in Canada are in a process of addressing the *Final Report of the Truth and Reconciliation Commission of Canada*.[6] Although there is much reason for scepticism, there are hopeful signs that Indigenous knowledge and perspectives are being welcomed as universities pursue an agenda of "Indigenization."

While important changes are in progress, Indigenous thinkers have always been aware of both the power and the contradictions that accompany participation in formal education systems shaped by Indigenous knowledge. In *Red Earth, White Lies*, Vine Deloria argues that the primary responsibility of "educated people" is to bring wisdom back into communities, where it can be examined and applied to the betterment of people as needed.[7] In 2002, Marie Battiste, Lynne Bell, and L.M. Findlay, reflecting on several decades of work to increase educational access and equity, observed that Indigenous knowledge remained marginalized within academic spaces.[8] Thus, there is a growing recognition of the limits to "decolonizing" the university as a Western institution. Instead, many Indigenous scholars emphasize the importance of pairing university-based education with community-led and land-based learning.[9] For instance, Dechinta, a community-led initiative, has become a model for land-based learning, and now offers programming in cooperation with several universities. In many ways, land-based learning initiatives bring the connection between activism and education full circle, and echo the spirit of the Native North American Travelling College, founded in 1969, which was ignited by a desire to revitalize Kanienkehaka language and culture.

Collectively, the intellectual and embodied energy we are seeing on campuses, and reflected in this edited collection, point to the possibility that "a third university is possible."[10] For many, a new university is necessary if the institution is to continue in its assumed role as a space for the formation and debate of ideas, values, and self-understandings that honour, respect, and centre Indigenous knowledges. Pursuing a hopeful path, and reflecting the spirit of this edited collection as a whole, this short essay asks two related questions: What theories and praxis do student activists bring with them? And what intellectual orientations, tools, and explanatory strategies do we see alive in student efforts to transcend the university-in-crisis and replace it with a better university, an anti-colonial university committed to broader systemic social and environmental change? In focusing on academic discussions, we keenly recognize that animating ideas herein come from Indigenous knowledge keepers, elders, animals, activists, and deeply personal histories. Our aim is to amplify these sites of learning. We seek to identify and explore a differently shared space, one where many communities, including those that make up this book (perhaps including you, the reader), encounter one another. The university is the setting that structures our conversation as an authorial collective: a professor of Anishinaabe and settler ancestry (Dorries) and two non-Indigenous settler researchers (Ritts and Hoogeveeen) who share disciplinary training in geography and working experiences within several Canadian institutions and communities. Below, we introduce four key intellectual coordinates – (1) critical theory; (2) environmental knowledges; (3) Indigenous legal traditions; and (4) creative methodologies – giving shape to the student voices collected in *The RAVEN Essays*. By no means comprehensive, they nevertheless help us map out some of the insurgent formations the student-led university embodies today, and from which better universities, and better worlds, might be created.

Critical Theory

Critical theory offers frameworks that can support – and not deflect from – decolonizing political and intellectual praxis.[11] By "critical theory" we mean a heterogeneous body of intellectual traditions – including critical Indigenous studies, Marxism, political economy, feminism, post-structuralism, and critical race theory – descended from, and reflexively opposed to, the dominant and assumed structures of Western society: class, race, hetero-patriarchy, gender violence, and nature. Above all else, critical theory has provided Indigenous thinkers with intellectual tools for uncovering the settler colonialism that imbues questions of class, race, and gender with their material and discursive force across the Indigenous lands known today in settler-colonial terms as North America.[12] Indigenous studies scholars have made crucial contributions to critical theory. For example, Glen Coulthard combines Fanon's account of racialization with Marx's notion of primitive accumulation to provide a critique of the "recognition"-based discourses that structure Canadian liberal multiculturalism.[13] Rob Nichols uses critical race theory to expound upon the assumed hierarchies that structure the apprehension of Indigenous bodies,[14] whereas Paul Nadasdy uses post-structuralist theory to observe the "subtle extensions of empire" mediated by the state's duty-to-consult process.[15] Through a deeply informed challenge to liberal contract theory, Audra Simpson uncovers the heroic, masculinist, and teleological fieldwork practices that continue to inform pedagogic models of working "on" Indigenous communities and knowledges.[16]

One compelling engagement reflected in *The RAVEN Essays* pertains to critical theoretical feminisms. Lisa Kahaleole Hall identifies "Indigenous feminism" as a project that "grapples with the ways patriarchal colonialism has been internalized within Indigenous communities as well as with analyzing the sexual and gendered nature of the process of colonization."[17] Here, feminism remains infused with critical theoretical debates around race, sexuality, class, and gender while focusing attention on the dichotomous categories (male/female; human/animal) that structure normative identities and enact gender-based violence on Indigenous bodies. This work is broadly epistemological in its concern with the role of dominant texts, policies, and representations – insisting, as Linda Tuhiwai Smith writes, on the persistence of "imperialism as a discursive field of knowledge."[18] But so, too, does it draw from formative Western philosophical ideas around existence and being. Notable here are powerful accounts of Indigeneity, in all its particularities, which scholars communicate by reworking Western conceptions of ontology with Indigenous intellectual traditions.[19]

Environmental Knowledges

Anishinaabe scholar Deb McGregor has made significant contributions to research in Indigenous environmental and social justice, continuing a tradition of Indigenous environmental thought that includes scholarship ignited by thinkers such as Gregory Cajete, Leroy Little Bear, and Willie Ermine. Ranging from interventions related to environmental regulation to work on "justice," McGregor's research furthers decolonial environmental thought and regulation by centring Indigenous environmental justice, which, she writes, is central to "all our relations," and outlines the beneficial relationships between all living things.[20] Similarly, Kyle Whyte draws on Anishinaabe intellectual traditions to counterbalance what he refers to as the ecological domination of settler colonialism and the subsequent violent undermining of Indigenous social and self-determining resilience.[21] There has been a critical turn in environmental justice and political ecology research towards studies of kinship and responsibility. This welcomed shift follows decades of geographical knowledge that ignored Indigenous scholarship, and which proposed the environment as existing outside of the human, creating a common Western false dichotomy (e.g., human vs. nature) that Indigenous scholarship and thinking has meaningfully deconstructed. Carroll's work, for example, brings together tribal parks with political ecology and circles back to Indigenous-led ecological thinking. Works like the *Roots of Renewal* provide motivation for a new generation of environmental scholarship grounded in Indigenous self-determination.[22]

Today, scholars like Carroll, Whyte, Daigle, and McGregor continue to the push the boundaries of Indigenous environmental knowledge, demonstrating its wide applicability and its ability to deconstruct Western academic traditions that often uphold the nature/culture dualism and erases Indigeneity. Indigenous environmental knowledges emphasize the holistic nature of Indigenous environmental thought, showing how it is embedded in cultural practices, language, and relations with the natural world. As a set of dynamic, situated, life-sustaining relations between place and knowledge, such knowledge is suggestive of a political-ethical framework that demands a fundamental realignment of relations between the human and more-than-human worlds. Such interventions are particularly welcomed in the face of climate change. Indigenous policy reports from the United States that bring intersectional analysis together with Indigenous-informed community knowledge on the impacts of climate change in community are significant,[23] as greenhouse gas emissions form the source of political tensions in a far from even world where the dispossession of Indigenous lands is embedded in the erasure of Indigenous ways of knowing. This collection counters this erasure through the amplification of emerging scholars who bring Indigenous knowledge and legal orders to the fore.

Indigenous Legal Traditions

Val Napoleon, Hadely Freidland, and volume co-editor John Borrows, among others, have helped pave the way for a new generation of Indigenous legal work, including through the creation of the University of Victoria's Joint Degree Program in Canadian Common Law and Indigenous Legal Orders, which continues to bring Indigenous legal traditions into dynamic engagement with the precepts of Western law. Indigenous legal traditions are motivated by diverse articulations of healing, reciprocity, and grounded respect, which resonate at a time when the Truth and Reconciliation Commission's Calls to Action showcase the pressing need to close the education and employment gaps between Indigenous and non-Indigenous populations in Canada. This includes Call to Action 10(i), to provide funding to close such gaps within one generation, and 10(iii), to develop culturally appropriate curricula, among other key Calls to Action, like 10(iv), to protect and teach Indigenous languages for credit.[24] In support of these calls, this book privileges scholarship on Indigenous legal orders, including oral histories, stories, spiritual beliefs, and customary law.

The following chapters approach Indigenous legal orders and traditional governance systems and work to decolonize Western conceptions of what constitutes law and governance. For example, Da Chen writes about the environmental obligations in the Dish with One Spoon treaty, while others, including Danette Jubinville and Wade Houle, write of Indigenous legal orders in contrast to the *Indian Act.* Indigenous notions of kinship and relationality are two modes of thinking that also push back on Western legal systems. The theme of revitalizing Indigenous legal orders and redefining Indigenous law is featured throughout this collection.

Creative Methodologies

The fourth and final pole we highlight as significant to the collection is creative methodologies, which highlights the rich voices, presentation, and artful communication of ideas. Characteristic of the diverse methodologies featured in this book, these show a rejection of colonial stereotypes and presentation formats. Through this collection, we are presented with new ways of thinking about how to more effectively express Indigenous relations and relational forms of thought. Exemplifying creative methodologies as a theme in Indigenous scholarship, Leanne Simpson[25] and Kyle Whyte[26] fuse Anishinaabe (Neshnabé) intellectual traditions with contemporary forms of creativity, sociality, and self-determination (poetry and creative readings of science fiction, respectively). Candis Callison looks to contemporary social media (such as X, formerly Twitter) to explore how young Indigenous activists build relationships and form alliances across space and place.[27] Another example is the Kino-nda-niimi Collective's *The Winter We Danced* – an inspiration for this project – which observes all the ways Idle No More's social significance can be expanded if we consider song, poetry, essay, and dialogue alongside exegesis and "standard" narrative history.[28] Speaking animals, alternate realities, multiverses, and pluri-temporalities are hallmarks of Indigenous storytelling traditions, notes Grace Dillon.[29] This has, in different ways, inspired rich writerly experiments in attending to significant others, addressing obligations, and paying respect – as is evident in the works of Zoe Todd[30] and Vanessa Watts,[31] among others.

Methodological experiment, rediscovery, and re-presentation offer approaches to research that conventional knowledge formats tend to foreclose. As Dan Wildcat explains, in North America many Indigenous traditions tell us that reality is more than just facts and figures collected so that humankind might widely use resources.[32] Rather, to know "'it' – reality – requires respect for the relationships and relatives that constitute the complex web of life." In ways that resonate across the literatures referenced, methodological innovation addresses urgent questions of environmental reparation and renewal of Indigenous lands and cultures. If "colonial thought brought us climate change," as Leanne Simpson remarks,[33] then new modes of thinking will be needed to undo past harms and forms of violence. Against conventions that would separate Indigenous (and allied) artists, musicians, and storytellers from Indigenous academics who write papers about these issues, Simpson[34] – along with Jarrett Martineau,[35] Mishuana Goeman,[36] and others – seeks to re-situate Indigenous methodologies (and associated creativities) within classrooms, public spaces, music venues, and art galleries. An explosion in creative methodological approaches is now visible in many university humanities and social sciences programs,[37] animated by broader academic-institutional turns towards community-based research, digital storytelling, and mixed-methods research. This activity is rooted in Indigenous knowledges and responds to the theft of First Nations stories, customs, and knowledge wrought by settler colonialism and systemic institutional barriers. Long before the current moment of reconciliation, Marie Battiste and others were calling for the decolonization of education and the meaningful enactment of Indigenous pedagogies within the academy.[38]

The Path Ahead

The next chapter in this volume comprises an introductory essay by co-editor John Borrows entitled "Making Meaning: Indigenous Legal Education and Student Action." Borrows writes about the key role that students play in advancing knowledge and taking action in universities. His essay challenges us to contemplate life's purpose through the lens of an optimistic university. Borrows's autobiographical piece considers the exclusion of Indigenous communities and students from mainstream law programs, and the creation of space for Indigenous peoples and communities in academic settings through research, advocacy, and action.

Following this essay are eleven student prize-winning essays, arranged into three sections. The first essays are organized under the theme of "Principles." Part 1 starts with Danette Jubinville's powerful piece on *Indian Act* enfranchisement, gender equity, and kinship. Thematically, this section is rooted in Indigenous laws and ways of knowing that supersede Crown governments and colonial order. Indigenous principles of naming and knowing are highlighted, including in Da Chen's analysis of the Dish with One Spoon treaty, followed by Wade Houle's chapter on naming and belonging. Part 2, entitled "Relations," includes essays grouped according to the themes of land, fish, and food security. Laura Peterson begins this section with an essay on the cultural landscape of Wood Buffalo National Park, and the grounded Indigenous knowledges that subtend the appearance of a Canadian national park. Atlanta Grant follows with a broad discussion on the politics of food security across Turtle Island, while the subsequent piece by Erica Isomura explores stories relating to capitalism, colonialism, Indigenous stewardship models, and fish. Part 3, entitled "Struggles" – and the final substantive section of this collection – begins with André Bessette's essay on the historical resilience of the Nuu-chah-nulth world view in the face of colonial impositions, and is followed by Kevin Ly's examination of contemporary Indigenous struggles over water in British Columbia. Tosin Fatoyinbo then provides a fruitful look at the politics of recognition in the context of Indigenous legal systems, followed by the final student essay on the theme of struggle, Helena Arbuckle's look at colonialism and the Dakota Access Pipeline. Finally, the book concludes with a reflection by key RAVEN essay award judges in a penultimate roundtable. Featuring adjudicators spanning different years of the award, the judges' roundtable reflects on the student essays and the role of students and Indigenous activism inside and outside of the academy. Lastly, Sarah Hunt leaves us with a meaningful afterword. The book thus proceeds from an introductory piece by John Borrows that frames Indigenous student movements, followed by student essays organized thematically around three primary sections, and concludes with a reflective roundtable presented in interview format followed by a final reflection.

While not exhaustive, we hope in this chapter to have provided a useful introduction to the RAVEN essays and the threads that connect them. Together, these pieces mark the arrival of a new generation of Indigenous intellectuals into the academy over the ten-year period that these student essays span. We also hope to showcase work from emerging scholars and to reflect on academia and the interdisciplinary research that is strengthened through emergent Indigenous scholarship and leadership that continues to amplify student-driven movements into the future.

NOTES

1 Linda Smith, *Decolonizing Methodologies* (London: Zed Books, 1999), 1.
2 Audra Simpson and Andrea Smith, *Theorizing Native Studies* (Durham, NC: Duke University Press, 2014), 3.
3 R.A. Warrior, "Intellectual Sovereignty and the Struggle for an American Indian Future," *Wicazo Sa Review* 8, no. 2 (1992): 1–20.
4 David Martínez, *Life of the Indigenous Mind: Vine Deloria Jr. and the Birth of the Red Power Movement* (Lincoln: University of Nebraska Press and the American Philosophical Society, 2019).
5 Lalaine Alindogan and Tina Yong, "Student Groups Criticize UBC's Continued Investment in Mining, Oil, Weapons Amid Move to Divestment," *The Ubyssey*, 4 March 2021, https://www.ubyssey.ca/news/endowment-data -2020-ubc/. Reportage on campus activities, such as that featured in this article, demonstrate this.
6 *Truth and Reconciliation Commission of Canada: Calls to Action* (Winnipeg: Truth and Reconciliation Commission of Canada, 2015).
7 Vine Deloria Jr., *Red Earth, White Lies: Native Americans and the Myth of Scientific Fact* (New York: Scribner, 1995).
8 Marie Battiste, Lynne Bell, and L.M. Findlay, "Decolonizing Education in Canadian Universities: An Interdisciplinary, International, Indigenous Research Project," *Canadian Journal of Native Education* 26, no. 2 (2002): 82–95.
9 Kate McCoy, Eve Tuck, and Marcia McKenzie, eds., *Land Education: Rethinking Pedagogies of Place from Indigenous, Postcolonial, and Decolonizing Perspectives* (Abingdon, UK: Routledge, 2017).
10 la paperson, *A Third University is Possible* (Minneapolis: University of Minnesota Press, 2017).

11 Winona LaDuke, *All Our Relations: Native Struggles for Land and Life* (Boston: South End Press, 1999); Simpson and Smith, *Theorizing Native Studies*.

12 Settler-colonial condition extends globally, as many North America–based scholars recognize.

13 Glen Sean Coulthard, *Red Skin, White Masks: Rejecting the Colonial Politics of Recognition* (Minneapolis: University of Minnesota Press, 2014).

14 Robert Nichols, "The Colonialism of Incarceration," *Radical Philosophy Review* 17, no. 2 (2014): 435–55.

15 Paul Nadasdy, "The Anti-politics of TEK: The Institutionalization of Co-management Discourse and Practice," *Anthropologica* 47, no. 2 (2005): 228.

16 Audra Simpson, *Mohawk Interruptus: Political Life across the Borders of Settler States* (Durham, NC: Duke University Press, 2014); Audra Simpson, "'Tell Me Why, Why, Why': A Critical Commentary on the Visuality of Settler Expectation," *Visual Anthropology Review* 34, no. 1 (2018): 60–6.

17 Lisa Kahaleole Hall, "Navigating Our Own 'Sea of Islands': Remapping a Theoretical Space for Hawaiian Women and Indigenous Feminism," *Wicazo Sa Review* 24, no. 2 (2009): 16; Lisa Kahaleole Hall, "Strategies of Erasure: US Colonialism and Native Hawaiian Feminism," *American Quarterly* 60, no. 2 (2008): 278.

18 Linda Tuhiwai Smith, *Decolonizing Methodologies: Research and Indigenous Peoples*, 2nd ed. (London: Zed Books, 2012), 22.

19 Mario Blaser, "Ontology and Indigeneity: On the Political Ontology of Heterogeneous Assemblages," *Cultural Geographies* 21, no. 1 (2014): 49–58; Emilie Cameron, Sarah de Leeuw, and Caroline Desbiens, "Indigeneity and Ontology," *Cultural Geographies* 21, no. 1 (2014): 19–26; Sarah Hunt, "Ontologies of Indigeneity: The Politics of Embodying a Concept," *Cultural Geographies* 21, no. 1 (2014): 27–32.

20 Deborah McGregor, "Mino-Mnaamodzawin: Achieving Indigenous Environmental Justice in Canada," *Environment and Society* 9, no. 1 (2018): 7–24.

21 Kyle Whyte, "Settler Colonialism, Ecology, and Environmental Injustice," *Environment and Society* 9, no. 1 (2018), 125–44.

22 Clint Carroll, *Roots of Our Renewal: Ethnobotany and Cherokee Environmental Governance* (Minneapolis: University of Minnesota Press, 2015).

23 Kirsten Vinyeta, Kyle Powys Whyte, and Kathy Lynn, *Climate Change through an Intersectional Lens: Gendered Vulnerability and Resilience in Indigenous Communities in the United States* (Portland, OR: US Department of Agriculture, Forest Service, Pacific Northwest Research Station, 2015), 72.

24 *Truth and Reconciliation Commission of Canada: Calls to Action* (Winnipeg: Truth and Reconciliation Commission of Canada, 2015).

25 Leanne Betasamosake Simpson, *This Accident of Being Lost* (Toronto: House of Anansi Press, 2017).
26 Whyte, "Settler Colonialism."
27 Candis Callison, "Communal Matters and Scientific Facts: Making Sense of Climate Change," in *Popular Culture and the Civic Imagination: A Casebook*, ed. Henry Jenkins, Gabriel Peters-Lazaro, and Sangita Shresthova (New York: NYU Press, 2020), 223–30.
28 Kino-nda-niimi Collective, *The Winter We Danced* (Winnipeg: ARP Books, 2014).
29 Grace L. Dillon, "Indigenous Futurisms, Bimaashi Biidaas Mose, Flying and Walking towards You," *Extrapolation* 57, nos. 1–2 (2016): 345.
30 Zoe Todd, "An Indigenous Feminist's Take on the Ontological Turn: 'Ontology' Is Just Another Word for Colonialism," *Journal of Historical Sociology* 29, no. 1 (2016): 4–22.
31 Vanessa Watts, "Indigenous Place-Thought and Agency amongst Humans and Non-humans (First Woman and Sky Woman Go on a European World Tour!)," *Decolonization: Indigeneity, Education and Society* 2, no. 1 (2013): 20–34.
32 Daniel R. Wildcat, *Red Alert! Saving the Planet with Indigenous Knowledge* (Wheat Ridge, CO: Fulcrum Publishing, 2009), xi.
33 Leanne Simpson, "Dancing the World into Being: A Conversation with Idle No More's Leanne Simpson," interview by Naomi Klein, *Yes! Solutions Journalism*, 6 March 2013.
34 Leanne Betasamosake Simpson, *As We Have Always Done* (Minneapolis: University of Minnesota Press, 2017).
35 Jarrett Martineau, "Fugitive Indigeneity: Reclaiming the Terrain of Decolonial Struggle through Indigenous Art," *Decolonization: Indigeneity, Education and Society* 3, no. 1 (2014): i–xii.
36 Mishuana Goeman, *Mark My Words: Native Women Mapping Our Nations* (Minneapolis: University of Minnesota Press, 2013).
37 Ethnohistory Field School, UVic, accessed 4 November 2023, https://web .uvic.ca/stolo/reports.php/gallery/pdf/Marsh_Boxing_FieldSchool_2015 .pdf. This is an example of institutional embrace of alternative methods.
38 Marie Battiste, "Enabling the Autumn Seed: Toward a Decolonized Approach to Aboriginal Knowledge, Language, and Education," *Canadian Journal of Native Education* 22, no. 1 (1998): 16–27.

Chapter Two

Making Meaning: Indigenous Legal Education and Student Action

JOHN BORROWS

Introduction: Contextualizing the Big Questions

I struggle to find life's meaning. The book *The Hitchhiker's Guide to the Galaxy* says that the answer is 42. While I like this answer, this issue is not solely numeric or metaphysical. Ordinary questions about life's physical trajectory can be difficult, such as "Where did I come from?," "Why am I here?," and "Where am I going?" For example, try answering these queries with reference to your own educational, employment, relational, political, social, or economic status. You will find many strands in your answers that complicate understanding since causation and trajectories are not linear. Furthermore, some things simply cannot be known.

At the same time, I should not overstate these challenges. These questions may yield obvious responses when considering specific issues. This is particularly true regarding education. For example, people go to university to prepare for future employment. In contemplating the question "Where do I come from?," a university student might say that they "came from" a secondary school. In discussing "why they are here," they could talk about developing skills, discipline, and friendships while receiving academic credentials. Finally, successful students might report that "they are going" to use these skills to support themselves and their loved ones. Thus, it is possible to provide some tangible, tentative answers to life's biggest questions if we consider more modest, proximate ends.

Nevertheless, these enquiries become more difficult when considering our interior views, particularly those related to time frames beyond our lifetimes. Our emotional, psychological, intellectual, and spiritual lives are notoriously complex. Our internal ideas and feelings constantly shift. Furthermore, approaches to life's so-called meaning are influenced by our class, race, gender, culture, religion, age, sexual orientation, and other varied contexts. Our perspectives are also intergenerational. This gives rise to many perspectives on where, why, and how questions.

This essay examines how university students change understandings of where we came from, why we are here, and where we are going in relation to Indigenous and Canadian law. Part 1 examines more general student approaches to education, including the intrinsic and instrumental motivations of students and how those students influence their professors' teaching and writing. Part 2 highlights Indigenous legal education initiatives at the University of British Columbia, Osgoode Hall Law School at York University, the University of Toronto, and the University of Victoria. This second part shows that law students are a significant force in persuading law schools to add Indigenous courses, moot courts, clinical experiences, internships, land-based events, a scholarly journal, and teaching in legal traditions to the mix of learning opportunities in these institutions. Part 3 discusses the importance of graduate students in asking where we came from, why we are here, and where we are going in legal, political, social, and economic terms. Finally, part 4 discusses the development of the Joint Degree Program in Canadian Common Law and Indigenous Legal Orders (JID/JD) at the University of Victoria Law School as a dimension of student action in changing how we teach and practise law in Canada.

Ultimately, this essay concludes by suggesting that life's meaning, in part, revolves around our capacities for growth. This growth is enhanced when students compare, contrast, consider, criticize, concur, and reject other people's views, including those of Canada's legislatures and courts, as well as their own communities', professors', and colleagues' views. In following these paths, this essay concludes that it is not easy to find life's meaning, in either proximate or absolute terms. Nevertheless, striving to understand life's purposes, while accepting the task's possible futility, creates constructive tensions that appropriately challenge self-assured stories about where we came from, why we are here, and where we are going as individuals and societies.

Part 1: University Students and Proximate Life Purposes

Despite certain limits,[1] students ask big questions and reflect on life's purposes in universities. For example, university students see their own lives in broader contexts when they learn more about themselves, their neighbours, and people they will never meet. They pursue post-secondary education to grapple with perspectives very different from their own. In the process, students seek to better understand their communities' histories and contemporary circumstances, and to contribute to the larger world.

Other students study for reasons that are not obvious at the outset, particularly when they acknowledge that they do not really know themselves. These students go to school to learn who they may be or become. In other words, they attend university for intrinsic reasons. They are prepared to be surprised about their own past, present, and future. From these vantage points, asking students where they came from, why they are here, and where they are going may offer them a way to better understand precisely what they hope to forge and learn in university.

Many of these same students use their education to challenge existing conditions and fight injustice. They want to transform universities by bringing Indigenous normativity into the heart of the academy's operation and existence. They do not merely want to repair, reconcile, or engender good feelings, particularly if these actions would lead to further marginalization. They want to overturn structural limitations experienced by Indigenous peoples, which, ironically, are often reproduced in the very name of so-called reconciliation between Indigenous people and others. These points are illustrated by the award-winning papers reproduced in this book. Each chapter is motivated by an author's acceptance, alteration, transformation, and/or rejection of their prior views. You'll see that the search for environmental health in relation to Indigenous peoples is a theme in this book. Each essay demonstrates how university students are implicitly tackling larger pressing questions related to where they came from, why they are here, and where they are going.

Yet other university students are more cautious in their approaches to university life. Their interests are more instrumental. They go to school to confirm their own ideas, identities, feelings, and frameworks. They are no different than most of us in this regard. We all struggle with complexity. This is why humans use mental shortcuts, called heuristics, in identifying and assessing information. This favours the status quo and introduces bias into our decision making.[2] Heuristics often provide shortcuts for taking action because we do not have the time, information, or abilities to explore all the varied implications of the information that we receive.[3] Life's ambiguities can be overwhelming and thus, when learning, we necessarily use rough estimates about how to proceed when presented with new ideas. This happens in university even as we try to widen our horizons or opportunities. Default positions, which affirm our views, help us take action without having to analyse every new detail that we encounter. Thus, heuristics can serve us well.

Unfortunately, danger arises when our biases hide our preconceptions, predilections, and predispositions, particularly from ourselves. This can be a challenge for university professors and students if our identity-based frameworks prevent us from facing facts or respecting others' opinions. Thus, universities function best when they bring our own biases to our attention. The following examples demonstrate students' humility and resolve to both respect and challenge received wisdom. These examples show why students are at the heart of any university's mission because of how they reflect upon preconceived ideas and forge new pathways.

Speaking personally, I have been impressed with how university students, as a group, broaden our horizons and expand our views. I see more of how other people frame life's purposes because of their participation in class, office visits, faculty governance, and research output. My most important experiences in universities are with students. This is not to diminish professors' contributions. I merely acknowledge an obvious point: professors are not the fount of all knowledge; they are not necessarily the most innovative part of our post-secondary education systems. Students ask questions that often propel a professor's own work. Students' research advances knowledge in key ways as professors learn from, with, and alongside students as we teach, mentor, supervise, and read their work. As I said, they have been some of my greatest teachers. I absorb as much from them as any other source.

One reason student engagement can be enriching is because student experiences can be more varied when compared to those of their professors. Students therefore bring unexpected perspectives to their interactions. They teach faculty and one another in direct and vicarious ways. Some students read widely and pass along what they have learned in class, conversations, and essays. Moreover, life is not static. Through thirty years of teaching, I have noticed that each generation raises different concerns. Students ask tough questions about things that I rarely or unfairly consider. Since their collective life experience is broader than my own, I would be foolish not to learn from them. Not only do they answer questions, but they also question answers that I assumed were not subject to further interrogation.[4] When this happens, I learn more about life's meaning.

There are also times when students teach me what not to do. Like all people, they can be foolish. Their missteps and oversights give me pause. They also make mistakes; they get things wrong and require correction. Like their professors, they make errors. How we convey and respond to sub-par grades, teaching evaluations, and peer review can be done well or poorly. Furthermore, a lapse in judgment can reveal assumptions that would otherwise remain hidden.

Furthermore, every decade, one or two students unfortunately cause great harm to their colleagues and/or our school. Whether through plagiarism, passive acquiescence, misdirected activism, intimidation, harassment, or obliviousness, a few students have torn us apart in detrimental ways. I have seen professors do the same thing. These are tough moments. We do all that we can to prevent them. Yet even these terrible experiences also generate lessons; they are teachable moments of life's purposes.

Fortunately, 99 per cent of my interactions with students have been enriching. Their essential role requires continual highlighting: universities would be significantly diminished if they were solely research labs or partisan think tanks. If students did not keep posing big questions, our enquiries would lose valuable context and content. In this way, they help us deepen searches for life's meaning.

Part 2: Law Students and Indigenous Legal Education Initiatives

While individual agency is key in considering life's big questions, we must also pose our queries collectively. We need one another's insights. We live in communities that require one another's care and contrariness. We are warmed by fires that we have not built, and we drink from wells that we have not dug. We are entangled.[5] What we do has an impact on those around us. Our actions have consequences for people whom we will never know. Individuals are not isolated atoms living in a social vacuum. For this reason, in addition to strongly encouraging individual freedom to question in unique ways, we must also approach life's purposes as groups.

In this light, students are also key to universities because they do more than educate us as individuals. They change how we collectively organize and communicate knowledge. In this way, they are an important part of how universities govern themselves. Students bring systemic concerns to the fore. They have an enormous impact on how we structure learning, including what we teach and write. This is particularly the case with Indigenous issues.

For example, students have greatly influenced universities' work in Indigenous law through the years. While they have not uncovered where we came from, why we are here, and where we are going from metaphysical standpoints, they have substantially reframed approaches to these questions from historical, political, economic, ecological, and social perspectives. Their demand for relevant curricula, support, and community-based learning has changed how and what we learn. Societally, while myriad problems remain, many universities now more clearly acknowledge that we come from a colonial past, and they must grapple with its ongoing effects to enhance our future health and well-being. These are significant "where, why, and what-for" contributions that enhance searches for life's meaning.

To illustrate, I will provide a few examples of how I have seen students shape the way in which we collectively ask questions in law schools, drawn from my own experience.

In 1992, when I taught at the University of British Columbia (UBC) Law School, I was director of the First Nations Legal Studies Program.[6] I chaired a First Nations Legal Studies Committee that had a significant number of student representatives who later became leaders in their field.[7] The students were not satisfied with how the school prepared them for practice or how we reflected their concerns. As a result, they requested more classes in both Indigenous peoples' own law and Canadian law as it relates to Indigenous peoples. Their views resulted in an increase in courses from two to seven and led to more people hired to teach in the field.

Furthermore, UBC students wanted an Indigenous legal clinic, a national Indigenous moot court, and more opportunities to associate with lawyers who represented Indigenous people. Their study, advocacy, and support helped form the Indigenous legal clinic in Vancouver's Downtown Eastside.[8] Their efforts at UBC and elsewhere led to the creation of a nationwide Kawaskimhon Moot experience that continues to be part of national law school curricula over twenty-five years later.[9] Métis student Jean Teillet at the University of Toronto Law School was a key individual in generating this development.[10] While getting the profession to support Indigenous students has been more challenging, student calls for the profession's backing remains an important goal at all law schools.[11]

When I later taught at Osgoode Hall Law School at York University, Indigenous student leadership also animated my experience. They helped form a "why we were there" enquiry, which in turn informed "where they would go." As noted, Indigenous communities and students were not well served by law schools' conventional curricular offerings. Anishinaabe law student Susan Hare was instrumental in the establishment of the Indigenous Lands, Resources and Governments Intensive Law Program at Osgoode, where I was hired to teach in 1994. Ms. Hare worked with faculty members to design the program's major features and fundraise to support its operation. As the program began, and through the years, it has remained relevant as it continues to be responsive to student feedback.

The Osgoode Intensive Program has now been running for over twenty-five years and has enrolled over four hundred and fifty students from many law schools.[12] These students spend an entire semester working with Indigenous communities around the world, in places that the students often have an active role in choosing. They study in an academically rigorous, placement-based course that contains their entire credit load for the term. A series of intensive seminars early in the semester precedes seven-week field placements with Indigenous organizations, environmental organizations, Indian reserves, law firms, or government departments working on applied legal issues. Students' written work has made a significant difference to these groups. Their research papers ensure they stand back and critically assess their experiences in order to ask bigger where, why, and what-for questions.

Indigenous student leadership was also key to broadening the legal academy's horizons when I taught at the University of Toronto Law School. Planning began in the late 1990s, and the students' efforts eventually led to the establishment of the *Indigenous Law Journal* (*ILJ*) in 2002. Douglas Sanderson[13] and Karen Abbott[14] were key figures in its development. The *ILJ* is a student-run legal journal "dedicated to developing dialogue and scholarship in the field of Indigenous legal issues both in Canada and internationally. The journal publishes articles, notes, case comments, and reviews grounded in all areas of study pertaining to both the laws of indigenous peoples and the law as it affects indigenous peoples."[15] The *ILJ*'s publications throughout the past two decades have demonstrated how students make a significant contribution to scholarship during their university studies, and how they have invited others to do the same. As with other initiatives described herein, the *ILJ* encourages big questions, which often go to the heart of where we came from, why we are here, and where we are going in Indigenous and other legal relationships.

During my time at the University of Toronto, law students also assisted in developing the June Callwood Program in Aboriginal Law when a donor approached me seeking to support Indigenous law students. After conversations with Dean Ron Daniels and the Indigenous Law Students' Association, the Callwood Program was initiated and now provides up to two summer fellowships of up to $10,000 for law students to work with community organizations on Indigenous legal issues.[16]

In 2001, I was hired as the Law Foundation Professor in Indigenous Justice and Governance at the University of Victoria (UVic) Law School. In taking this position, I also saw how law students helped propel Indigenous law scholarship and program development. In so doing, they further changed how we approached "why we are here" and "where are we going" questions. For example, as Nunavut became a new territory, there was a need to provide legal education in the North because there was only one Inuit lawyer in the area when the territory was created.

UVic Law School co-op student Kelly Gallagher-McKay (who, a decade later, became a university professor[17]) worked with Inuit leaders, other leaders, and organizations in Iqaluit to develop a proposal to host legal education in the North. Through her efforts, as well as other student and community efforts, UVic ran the Akitsiraq Law School from 2001 to 2005 in partnership with the Akitsiraq Law Society and Arctic College in Iqaluit, Nunavut.[18] The program operated in the North on a cohort model with students proceeding together through a four-year law degree.[19] Most of the students were Inuit, and they taught me much about Inuit law along the way. Inuit law has metaphysical and practical orientations that expanded my understanding of life's meaning once again.

In particular, the students took an Inuit language and law course with Inuit Elder-in-Residence Lucien Ukaliannuk, which caused them to bring questions about Indigenous law more generally to my classes on contracts (voluntary obligations), remedies, and Indigenous peoples' and Canadian law.[20] In addition, the students asked questions about how to live well, how to become a balanced legal professional, and how to fulfil their responsibilities to past and future generations. They wanted to be responsively responsible to Inuit lifeways in how they would, one day, practise law. They were very proactive learners. The students' questions helped me see the place of Inuit law in Nunavut and their views reinforced the importance of Indigenous law more generally, as necessary to understanding law in Canada.

Law students from the South learned lessons about Indigenous law's importance too. Akitsiraq students met other law students in our Victoria-based UVic JD program in the Akitsiraq cohort's final year.[21] Inuit students enrolled in our conventional JD courses on the southern edge of Vancouver Island to complete their degree. In Victoria, they introduced their perspectives on law to faculty and other law students.

The broader community outside the law school also gained a heightened appreciation for Indigenous law when Inuit law students and elders met with Kwakwaka'wakw elders and law keepers in the Mungo Martin Big House next to Victoria's Inner Harbour. Around a blazing fire, surrounded by clan and crest carvings, Inuit and Kwakwaka'wakw people shared their laws. They explained how they regulated resource use, land allocation, family matters, and other contemporary issues. They talked about larger philosophical questions that placed us in a broader stream of time. We talked about the stars, winds, ice, rain, elements, and animals and how they all give rise to obligations. We discussed their millennia-long history, colonialism's recent impacts, and their future hopes for greater self-determination. This was a trans-systemic engagement between legal systems through story, song, dance, and discussion.

It revealed innovative approaches to where, why, and what-for questions. It was not long before a proposal was drafted for a joint Indigenous law and common law degree for UVic.

The idea of a joint Indigenous law and common law degree was first discussed in the early 1990s. Professors Trish Monture,[22] James (sákéj) Henderson,[23] my PhD student Harold Cardinal,[24] and I generated the idea. We were teaching in the Summer Program in Native Legal Studies at the Native Law Centre in Saskatoon, Saskatchewan, and the students wanted more Indigenous law content.[25] In response, our idea was to teach Indigenous peoples' own legal orders in law school settings, including creating a comprehensive program hosted by the First Nations University (FNU) in Regina. When the FNU law school idea failed to transpire, I became fixated on the idea of an Indigenous law degree. From this experience, and building on my graduate work,[26] my subsequent scholarship dove more deeply into Indigenous peoples' own law and its relationship with broader Canadian law.[27] Thus, in 2005, as Akitsiraq was winding down, I composed a thirty-page document proposing a joint Indigenous law and common law degree at the UVic Law School.[28] With the encouragement of our then law dean Andrew Petter,[29] Canada Research Chair Jeremy Webber,[30] and Jim Tully,[31] the proposal outlined the broader contours and specific details of a joint Indigenous law and common law degree.

Law students at UVic and elsewhere supported the necessity of seeing Indigenous law as an ongoing force in Canada. UVic students, led by Doug White, formed an Indigenous Law Club that provided excellent feedback on the proposal.[32] They generated further student and Indigenous community interest, and their enthusiasm led to the creation of an initiative that piloted different aspects of the future degree. Along the way, UVic offered summer programs in Indigenous law, specialized Indigenous law methods courses, and land-based experiences with communities, experiences in which students eagerly partook. Professor Val Napoleon was hired and planning for the joint Indigenous law and common law degree kicked into high gear.[33] Among other activities, Professor Napoleon created an Indigenous Law Research Unit to work with communities to help them revitalize their laws as they invited us into their work. Some of the initiatives were national, and students from across the country enrolled in them, which further enhanced the field.[34] Student class participation and research papers revealed other dimensions of the field, and these students took their experiences at UVic back to their law schools and communities.

This soon led the law school to work more closely with Indigenous communities. Before long, we were invited to Osgoode Hall Law

School,[35] McGill Law School,[36] Western University Law School,[37] Windsor Law School,[38] the University of Toronto Law School,[39] and others to offer land-based Indigenous law courses and orientations for even more students.[40] When you are out on the land, forests, fields, shorelines, and in the homes of Indigenous communities, questions about where we came from, why are we here, and where are we going inevitably arise. In our time together, elders, wisdom keepers, storytellers, chiefs, band councillors, and First Nations members both prompt and respond to these questions. The discussions are wide-ranging. Fortunately, when students interact with one another, Indigenous communities, and the natural world in these settings, you cannot contain the scope of conversation within conventional university bounds.

Part 3: Graduate Students and Indigenous Law

As Indigenous legal education efforts grew alongside more frequent attentiveness to life's broader questions, Indigenous law initiatives arose at many schools. This led to a number of Indigenous and other students enrolling at UVic Law School with the goal of studying Indigenous law in their graduate degree programs. They wanted to contribute to the development of the field in university and community settings. Many of these graduate students are now professors and teachers at other schools and some are colleagues at UVic, teaching in the school's JID/JD joint degree program. A summary of their research in the next paragraph demonstrates their work's groundbreaking reach. From their titles, you can also see that many asked questions that reframe where we came from, why we are here, and where we are going, at least legally speaking. Some of their work even poses such questions on a much wider scale and in an exceptionally deep way.

The research on Indigenous peoples' law and/or its more general relationship with Canadian law is burgeoning. It now influences countless law students, and others, as these former graduate students teach and continue to write through their professional careers. As noted, in the last ten years their questions gave rise the following research: Ryan Beaton, *Positivist and Pluralist Trends in Canadian Aboriginal Law: The Judicial Imagination and Performance of Sovereignty in Indigenous-state Relations*;[41] Kinwa Bluesky, *Art as My Kabeshinan of Indigenous Peoples*;[42] Andrée Boisselle, *Law's Hidden Canvas: Teasing Out the Threads of Coast Salish Legal Sensibility*;[43] Keith Cherry, *Practices of Pluralism: A Comparative Analysis of Trans-systemic Relationships in Europe and on Turtle Island*;[44] Robert Clifford, *WSÁNEĆ Law and the Fuel Spill at Goldstream*;[45] Aimée Craft, *Breathing Life Into the Stone Fort Treaty*;[46] Alan Hanna, *Dechen Ts'edilhtan: Implementing*

Tsilhqot'in Law for Watershed Governance;[47] Carwyn Jones, *The Treaty of Waitangi Settlement Process in Māori Legal History*;[48] Dawnis Kennedy, *Aboriginal Rights, Reconciliation and Respectful Relations*;[49] Darcy Lindberg, *Nêhiyaw Âskiy Wiyasiwêwina: Plains Cree Earth Law and Constitutional/Ecological Reconciliation*;[50] Danika Littlechild, *Transformation and Re-formation: First Nations and Water in Canada*;[51] Johnny Mack, *Thickening Totems and Thinning Imperialism*;[52] Aaron Mills, *Miinigowiziwin: All That Has Been Given for Living Well Together: One Vision of Anishinaabe Constitutionalism*;[53] Sarah Morales, *Snuw'uyulh: Fostering an Understanding of the Hul'qumi'num Legal Tradition*;[54] Val Napoleon, *Ayook: Gitksan Legal Order, Law, and Legal Theory*;[55] Joshua Nichols, *Reconciliation and the Foundations of Aboriginal law in Canada*;[56] Nicole O'Byrne, *Challenging the Liberal Order Framework: Natural Resources and Metis Policy in Alberta and Saskatchewan (1930–1948)*;[57] Jacinta Ruru, *Settling Indigenous Place: Reconciling Legal Fictions in Governing Canada and Aotearoa New Zealand's National Parks*;[58] Nancy Sandy, *Reviving Secwepemc Child Welfare Jurisdiction*;[59] Jennifer Sankey, *Globalization, Law and Indigenous Transnational Activism: The Possibilities and Limitations of Indigenous Advocacy at the WTO*;[60] and Kerry Sloan, *The Community Conundrum: Metis Critical Perspectives on the Application of R v Powley in British Columbia.*[61]

As you see, graduate students are making a significant difference in examining the resurgence of Indigenous law and its relationship to the Canadian state. Along the way, they reframe some of society's biggest questions. Reading their work reveals the unsettling nature of the state's framing of society's foundation, organization, and goals. They discuss how the assumptions built into Canadian legislation, case law, and governance accentuate acquisitiveness in ways that conceal Indigenous views of the state's origin, operation, and future fate.

In noting this work, I should also highlight that there are many other graduate students who write in this field. In the prior paragraph, I only referenced UVic graduate students who subsequently taught in university settings; the individuals above are all university professors. Other former UVic graduate students writing in this field work in key positions in governments, law firms, courts, Indigenous communities, and nongovernmental organizations. Furthermore, while UVic Law School has been fortunate to host a large number of students working with Indigenous legal issues, other law schools have also experienced a significant expansion of people working in this field.[62] This activity demonstrates that students are a leading force in assisting Indigenous communities in the communication and revitalization of Indigenous peoples' own laws. In the process, they help our country more clearly see where we came from, why we are here, and where we are going in relation to these fundamental relationships.

Part 4: Colonialism and the Resurgence of Indigenous Law: The JID/JD Joint Degree

Despite the positive changes outlined in this chapter, there are important questions that are less likely to receive consideration from other Canadian legal institutions. Courts and legislatures have consistently avoided big questions such as how Canada can claim legitimate governance over Indigenous peoples.[63]

For example, the Crown's claim to overarching governance rests on assumptions of Indigenous political, legal, and social inferiority when the parties encountered each other.[64] Crown claims arise solely from unilateral Crown assertions without any meaningful participation or engagement from Indigenous peoples who were "here first."[65] The Crown's assertions are a weak foundation from which to ask "where did we come from" questions.[66] One-sided, racist suppositions make it difficult for the Crown to admit that its claims of where it came from as government did not adequately deal with prior Indigenous land relationships and social orders.

Moreover, even if we could somehow legitimate the unjust sources of government power relative to Indigenous peoples, the next question is equally challenging: How can Canada justify governing Indigenous peoples when the contemporary effects on Indigenous peoples are disastrous? Indigenous peoples' life expectancy is ten to fifteen years shorter than that of other people in Canada;[67] infant mortality rates are two to four times higher;[68] on-reserve employment is over 25 per cent below the national rate;[69] median income is approximately 50 per cent less than non-Indigenous income;[70] approximately 57 per cent of First Nations young adults on-reserve have completed high school, compared to 89.2 per cent of non-Indigenous young adults off-reserve;[71] and over 40 per cent of reserve homes require major repairs with problems in plumbing, water access, and quality, as well as exposure to allergens and mould.[72] Furthermore, Canada unequally and chronically underfunds First Nations education, childcare, and social services when compared with the general population.[73]

The undermining of Indigenous governance in Canada, and replacing it with federal and provincial so-called jurisdiction, is not justifiable in light of the ongoing prejudice, ineffectiveness, and harm perpetrated by Canada and the provinces. Other questions follow: How can the Crown justify claiming underlying title to land throughout Canada? Or how can it claim exclusive use and possession in places where Indigenous traditional owners have not transferred such rights to them through treaties, agreements, or other means?[74]

Another question reasonably flows from these enquiries: How do Canadian courts justify their role in adjudicating Indigenous and Crown disputes when they are creatures of the state with which Indigenous peoples are in conflict?[75] Courts have not meaningfully asked where we came from, why we are here, and where we are going when relating to Indigenous peoples. This is to say that courts do not clearly acknowledge their colonial roots, their ongoing colonial supervision, or their suppression of meaningful Indigenous self-determination.

The situation could have been different. In 1982, existing Aboriginal and treaty rights were recognized and affirmed in section 35(1) of Canada's *Constitution Act, 1982*. The opportunity to place Indigenous governance on solid constitutional grounds has not occurred. Furthermore, Aboriginal title recognition has been slow to develop, and Crown title over Indigenous land has not been justified either.[76] This is despite over fifty cases interpreting section 35(1) decided by the Supreme Court of Canada. These cases are in addition to the thousands of cases heard in lower-level bodies across the land. While section 35(1) has generated noticeable changes, such as requiring the Crown to justify its actions when breaching Aboriginal and treaty rights,[77] Indigenous peoples have been unable to have their most important powers clearly endorsed, such as Indigenous governance and law.[78] Moreover, the courts have not adequately justified where their authority comes from, why they are here, or where they are going (except to say that they are here to stay[79]), relative to Indigenous peoples throughout the land.

Fortunately, Canadian courts and legislatures do not hold all of the power when considering the revitalization of Indigenous peoples' law. Indigenous peoples deepen their own legal traditions in ways that decentre the state.[80] They ask and create their own answers about life's purposes.[81] In the process, Indigenous peoples have revitalized their laws by bolstering potlatch and feast traditions on the West Coast,[82] expanding Guardian Watchperson programs throughout the country,[83] strengthening their internal constitutional relationships from coast to coast,[84] and using custom and declarations to protect their communities through a worldwide pandemic.[85]

Universities generally, and law schools more particularly, engage with Indigenous communities to ask larger questions. As noted in the last section, this prompts students to produce innovative scholarship aimed at understanding how to teach Indigenous legal orders.[86]

Furthermore, the Truth and Reconciliation Commission of Canada also prompted law schools to engage more fully with Indigenous law and Canadian law dealing with Aboriginal peoples.[87] The commission's twenty-eighth Call to Action states the following:

> We call upon law schools in Canada to require all law students to take a course in Aboriginal people and the law, which includes the history and

legacy of residential schools, the *United Nations Declaration on the Rights of Indigenous Peoples*, Treaties and Aboriginal rights, Indigenous law, and Aboriginal-Crown relations. This will require skills-based training in intercultural competency, conflict resolution, human rights, and anti-racism.[88]

At UVic Law School, these varied efforts paved the way for the launch of the JID/JD joint degree in 2018. The JID/JD is a trans-systemic law degree that teaches Indigenous law alongside common law. This four-year program compares and contrasts legal traditions to understand where the gaps, inconsistencies, and taken-for-granted assumptions are present in conventional law school teaching and practice. In the first year, students take five classes. There is a course in constitutionalism that teaches Anishinaabe *chi-naaknigewin* alongside Canadian federalism, the *Charter of Rights and Freedoms*, and Aboriginal and treaty rights. This is the course that I convene.[89] There are written constitutions developing throughout Anishinaabe territory in Ontario,[90] and we also discuss Anishinaabe constitutions, legislation, and tribal court decisions from Minnesota, Wisconsin, and Michigan, since Anishinaabe law is also practised in these jurisdictions.[91] Other first-year courses engage Cree legal traditions and Canadian criminal law, Gitksan law related to land and Canadian property law, legal research and writing, and law, legislation, and policy. In the second year, students learn Tsilhqot'in laws of voluntary obligations, which is taught trans-systemically with contract law. There is a course in Hulqumi'num (Salish) law of involuntary obligations, which also conveys Canadian tort law principles throughout second year. Students take other classes from the general curriculum in the remainder of their second year. In upper years, they take trans-systemic business associations, trans-systemic administrative law, and electives from the general curriculum. In addition, they must take two community-based courses for an entire semester, immersing them in a particular legal tradition. Field courses exist in Hulqumi'num law, Cree law, Kwakwaka'wakw law, and W̱SÁNEĆ law.

There are approximately twenty-five students admitted each year, and they come from all walks of life. A slight majority of students are Indigenous, but the program is designed to appeal to students from many different backgrounds. It is an honour to teach all of them.

In the program's first year, students faced high expectations. They studied under the media's glare. They were the first students in the world enrolled in a professional degree combining the study of Indigenous and common law. They also experienced the culmination of many years of piloted courses, projects, scholarships, and planning as their first year unfolded. There were more than a few uneasy moments as everyone worked through the newness of the program. Fortunately, the students were brilliant. They have written outstanding papers, made significant contributions to Indigenous communities through their summer employment and coursework, and are now graduating to fill coveted positions in communities, government, law firms, and other places where Indigenous law interacts with contemporary affairs. As you can see, these students have a strong proximate sense of where they came from, why they are at UVic, and where they are going in the short term.

The second-year cohort was equally as strong in their preparation and participation. The students also faced unique challenges. In the middle of their first year, nationwide protests broke out against the Coastal Gas-Link Pipeline and in support of Wet'suwet'en hereditary chiefs.[92] One of the epicentres of protest was on the steps of the British Columbia legislature, in Victoria.[93] Many students spent weeks camped out on the legislature's lawn to advance Indigenous law as a vehicle for addressing energy conflicts in the province and beyond.[94] There were also healthy differences of opinion among the student group. While these students definitely knew where they came from before entering UVic, and why they were studying at the school, the protests gave rise to further questions concerning where they might go with their education, given the conflict between Indigenous law and Canadian law more generally.

The third year of the program occurred during a worldwide lockdown due to the COVID-19 pandemic. As I prepared my course, I was concerned that we would not effectively learn Anishinaabe and Canadian constitutionalism because we could not meet face-to-face. I normally take my students outside when learning Anishinaabe law; this allows us to draw teachings, processes, and principles from the natural world. But I was worried that virtual learning would interfere with this process. When we are outside together, we might see a robin (*pitchii*), which would prompt discussion about freedom in Anishinaabe and Canadian constitutional traditions.[95] Anishinaabe people can "read" this bird as communicating liberty, choice, autonomy, and self-determination because of the narratives associated with it.[96] When learning outdoors, I could reference hundreds of "cases" from the world's living law library to illustrate legal relationships, processes, and principles. After a discussion of Anishinaabe laws from our oral traditions, I might then reference various written Anishinaabe constitutional freedoms before discussing Anishinaabe Tribal Court cases dealing with these responsibilities.[97] This would eventually lead to discussions of section 7 *Charter* rights to life, liberty, and security, as we compared and contrasted Supreme Court of Canada approaches with Anishinaabe constitutionalism on this point.[98] Finally, I might drum and sing an Anishinaabe song referencing freedom, or discuss the etymology of the word "freedom" in Anishinaabemowin, *dibe-nindizowin*, which can mean that a person possesses liberty within themselves and their relationships.[99]

Fortunately, the students embraced working together through the challenges of COVID-19, despite our distance and biweekly trans-systemic constitutional law classes held over Zoom, the videoconferencing software. With their enthusiasm and modifications to the curriculum, we learned and applied Anishinaabe and Canadian constitutionalisms in a revised format. Six months after the course was finished, the students presented me with a small book of reflections regarding what they learned during the year. In their short commentary, I was encouraged to see that Anishinaabemowin (the Ojibwe language) figured prominently. In regard to constitutional law, they discussed *debwewin* (truth),[100] *wazhash-koonh* (muskrat),[101] *giiwedanang* (the North Star),[102] *zaagi'idiwin* (love),[103] *aamoyag* (bees),[104] *manidoo makwa* (spirit bear song),[105] *michi-dawada-naa* (space or the great opening),[106] *akinoomagewin* (learning from the earth),[107] *nigig* (otter from clan teachers),[108] *memengwaag* (butterflies),[109] *aabawaawendam* (forgiveness),[110] *ningowaaso-miigiwewin* (seven gifts or ancestral teachings),[111] *bimaadiziwin* (life),[112] *zoongidewin* (courage),[113] and *nibwaakaawin* (wisdom).[114] Of course, the students also learned principles of federalism, Aboriginal and treaty rights, and *Charter* rights from both Anishinaabe and Canadian constitutional perspectives in our time together. However, the students' messages to me emphasized that they were thinking more broadly about the materials than often occurs in a conventional first-year constitutional law course. In fact, I hope that you can see from the Anishinaabe words above that the students were engaging with constitutional themes discussed in this essay: where we came from, why we are here, and where we are going. All of this once again demonstrates how students are taking significant steps in advancing the resurgence of Indigenous law.

When students learn about Indigenous legal traditions, they discover other ways of formulating and practising law. For instance, Indigenous peoples' law is often explicitly connected to their community aspirations, such commitments to a living earth, intergenerational care, and healing. These kinds of principles cause students to ask questions about life's potential purposes raised by these ambitions. Of course, Indigenous peoples (like all other legal communities) do not always live up to their highest aspirations, and this causes students to enquire about human frailty in relation to law too. Moreover, the students also learn about the challenges that Indigenous peoples encounter in meeting their goals within Canadian law because it marginalizes them. These concerns cause students to pose further questions about fairness and justice, whether life has any meaning, or whether they can discern its broader purposes.

Conclusion: Education, the Meaning of Life, and Indigenous Law

At the end of this essay, I return to the question with which this chapter began: What is the meaning of life? The ensuing discussion suggested that, if there is an answer, it revolves in part around our capacities for growth. Student initiatives, described in this essay, and in the chapters that follow, exemplify this pattern. As we grow, we enhance our abilities to discern our own life's purposes. As we learn, we gain insight about where we came from, why we are here, and where we are going. In this quest, we benefit from comparing, contrasting, considering, criticizing, concurring with, and rejecting other people's views. To illustrate this process, we have considered how law schools, and perhaps even broader society, benefit from understanding Indigenous perspectives that were previously marginalized and concealed. These examples suggest that discovering life's meaning is not solely an individual experience; it flows from the quality and sustainability of our relationships. This suggests that humans make their own meanings even as forces beyond our control also form our purposes.

If understanding life's meanings is simultaneously within and beyond our capacities, this implies that we should work hard to apprehend it, while also recognizing the potential impossibility of the aim. If this seems like a contradiction, it is a healthy one. This paradox facilitates creativity. Striving to understand life's purposes while accepting the task's possible futility pushes us into unknown territory. This creates a constructive tension that is key to learning.

Education would be regressive if its sole purpose was to confirm what we already know. It would lose its dynamic power if it merely involved the restatement of pre-existing truths. However, as this essay has shown, using Indigenous examples, education must go further and challenge our assumptions and practices. It must move beyond repetition or risk becoming an empty ritual. Replicating what we already know is necessary, but insufficient, to meet our own needs and society's demands.

Fortunately, student research, advocacy, negotiation, and hard work are widening our views about what at times seems impossible – respecting Indigenous peoples' search for life's purposes within and beyond Canada's legal frameworks. This work is slowly gaining momentum as students shed greater light on our collective past, present, and future. In the process, this impacts society as a whole, as student ideas and research spreads beyond universities through their work.[115]

Despite the immensity of the task, whether finding the meaning of life or changing Canada's legal structure as it relates to Indigenous peoples, there is great worth in tackling what seems impossible. Challenging the status quo can help us see more clearly where we came from, why we are here, and where we are going as individuals and society, even if we do not believe or appreciate the answers that we receive.

NOTES

1 John Borrows, *Law's Indigenous Ethics* (Toronto: University of Toronto Press, 2019). Law schools, which are the locus of enquiry in this essay, do not explicitly engage with ultimate, transcendent, or metaphysical questions. Their treatment of ultimate ends, goals, aims, and objectives is often subtle, indirect, and embedded deep within assumptions about the meaning and purpose of law in Western democracies. Perhaps law schools do not explicitly engage in "purpose of life" enquiries because there is a genuine concern that choosing, identifying, or following specific ends could devalue or ignore stories in reductive, exclusionary, and harmful ways. This is one reason why enquiries from religious, spiritual, emotional, and psychological perspectives are scarce in these settings. Furthermore, there is a strong view that perceiving ultimate ends is beyond human discernment and comprehension. Some also think that ideas about ultimate designs are false, fabricated, fake, and fictitious. Thus, very few law professors speak about God, gods, the sacred, or the inner complexities of teleological human motivations in classes. While these issues are not completely absent, they play a vanishingly small role in our discussions, despite how important they may be to individuals, groups, and entire societies outside of legal education. In fact, if these issues are raised at all, it is often (though not always) through their dismissal as being irrelevant to contemporary legal understanding and practice. As noted, some even consider that discussion of the spiritual or supernatural can be distracting and dangerous. Marginalization of these questions is revealed in how law schools structure class discussions and teaching materials. While I have views about the hidden nature of law school's teleological commitments (see generally the above reference), this essay focuses on more relative, proximate engagements with large questions about life's meaning.

2 Daniel Kahneman, *Thinking, Fast and Slow* (Toronto: Anchor, 2013). This is an excellent analysis of heuristics and biases.

3 Eyal Zamir and Doron Teichman, eds., *The Oxford Handbook of Behavioral Economics and the Law* (New York: Oxford University Press, 2014); Mark

Kelman, *The Heuristics Debate* (New York: Oxford University Press, 2011); Richard H. Thaler and Cass R. Sunstein, *Nudge: Improving Decisions about Health, Wealth, and Happiness* (New York: Penguin Books, 2009). These books include a discussion of heuristics and biases in a legal setting.

4 Sarah Noël Morales, "*Snuw'uyulh*: Fostering an Understanding of the Hul'qumi'num Legal Tradition" (PhD diss., University of Victoria, 2014), 42.

5 Borrows, *Law's Indigenous Ethics*, 115–20.

6 "Indigenous Legal Studies," Peter A. Allard School of Law, UBC, accessed 17 May 2022, perma.cc/AYW3-YSBH. The UBC Law School changed the name of the First Nations Legal Studies Program to the Indigenous Legal Studies Program. Information about the program can be found at the website above.

7 Diane Haynes, "Ardith Wal'petko We'dalx Walkem," Peter A. Allard School of Law, UBC, accessed 17 May 2022, perma.cc/R5RJ-B5PN; Dan Fumano, "'It's about Time': Leading First Nations Law Expert Named to B.C. Supreme Court," *Vancouver Sun*, 22 December 2020, perma.cc /A3B8-UV84; Ardith Walkem and Halie Bruce, eds., *Box of Treasures or Empty Box? Twenty Years of Section 35* (Penticton, BC: Theytus Books, 2003); "Terri-Lynn Williams-Davidson," Raven Calling Productions, accessed 24 September 2022, perma.cc/WP3U-66Q6; Diane Haynes, "Terri-Lynn Williams-Davidson" Peter A. Allard School of Law, UBC, accessed 17 May 2022, perma.cc/4XWN-RR2P; Gid7ahl-Gudsllaay Lalaxaaygans (Terri-Lynn Williams-Davidson), O*ut of Concealment: Female Supernatural Beings of Haida Gwaii* (Victoria, BC: Heritage House, 2017); "Darwin Hanna," Peter A. Allard School of Law, UBC, accessed 17 May 2022, perma.cc/BXJ8-P69K; Darwin Hanna and Mamie Henry, eds., *Our Tellings: Interior Salish Stories of the Nlha7kápmx People* (Vancouver: UBC Press, 1995); Sheldon Cardinal, "The Spirit and Intent of Treaty Eight: A Sagaw Eeniw Perspective" (LLM thesis, University of Saskatchewan, 2001); "Leona Sparrow," Peter A. Allard School of Law, UBC, 17 May 2022, perma.cc/ZML9-RS8V; Leona Sparrow, Jordan Wilson, and Susan Rowley, "ĆƏSNAʔƏM, The City before the City: A Conversation" *BC Studies* 45 (2018): 199; "Cynthia Callison," Peter A. Allard School of Law, UBC, accessed 17 May 2022, perma.cc/Q5BJ-2QC9; Cynthia Callison, "Appropriation of Aboriginal Oral Traditions," *UBC Law Review*, special issue (1995): 165–81; Penny Cholmondeley, "Duncan McCue," Peter A. Allard School of Law, UBC, accessed 17 May 2022, perma.cc/4YGD-ZFUE; Duncan McCue, *The Shoe Boy: A Trapline Memoir* (Vancouver: Purich Books, 2020). This is a small sample of Indigenous students who made a contribution to student government at this time, among others.

8 Patricia Barkaskas and Sarah Buhler, "Beyond Reconciliation: Decolonizing Clinical Legal Education," *Journal of Law and Social Policy* 26, no. 1 (2017):

7; Sarah Buhler, Chantelle Johnson, Nicholas Benkinsop, Leif Jensen, and Kim Pidskalny, "Clinical Legal Education on the Ground: A Conversation," *Journal of Law and Social Policy* 32 (2020): 127.

9 "Kawaskimhon Moot 2021," University of Saskatchewan College of Law, March 2021, perma.cc/3LCH-Z392; Lara Ulrich and David Gill, "The Tricksters Speak: Klooscap and Wesakechak, Indigenous Law, and the New Brunswick Land Use Negotiation," *McGill Law Journal* 61, no. 4 (2016): 979. Information about the most recent Kawaskimhon Moot can be accessed in the first reference. For an example of the kind of arguments made at the moot, see the Ulrich and Gill article.

10 "Jean Teillet," Pape Salter Teillet LLP, accessed 8 August 2022, perma. cc/763Y-JGG6; Jean Teillet, *The North-West Is Our Mother: The Story of Louis Riel's People, the Métis Nation* (Toronto: HarperCollins Canada, 2019). Jean Teillet's accomplishments since graduating from the University of Toronto Law School are listed on her law firm's webpage. For a sample of her writing, see *The North-West Is Our Mother*.

11 Sonia Lawrence and Signa Daum Shanks, "Indigenous Lawyers in Canada: Identity, Professionalization, Law," *Dalhousie Law Journal* 38, no. 2 (2015): 503. This includes a discussion of Indigenous peoples and the legal profession.

12 "Intensive Program in Indigenous Lands, Resources & Governments," Osgoode Hall Law School, York University, accessed 17 May 2022, perma. cc/8CY5-ZHQH. Information about the Intensive Program in Indigenous Lands, Resources, and Governments can be found here.

13 Douglas Sanderson (Amo Binashii), "The Residue of *Imperium*: Property and Sovereignty on Indigenous Lands," *University of Toronto Law Journal* 68, no. 3 (2018): 319; Douglas Sanderson (Amo Binashii) and Amitpal C. Singh, "Why Is Aboriginal Title Property if It Looks Like Sovereignty?," *Canadian Journal of Law and Jurisprudence* 34, no. 2 (2021): 417. Professor Sanderson is now a tenured professor at the University of Toronto Law School; the above references are a sample of his recent writing.

14 Karen L. Abbott, "Urban Aboriginal Women in British Columbia and the Impacts of Matrimonial Real Property Regime," in *Aboriginal Policy Research*, vol. 2, ed. Jerry P. White, Paul Maxim, and Dan Beavon (Toronto: Thompson Educational, 2004), 165. Karen Abbott is now a judge on the Provincial Court of British Columbia. She completed her PhD at Royal Roads University. The above reference is a sample of her legal writing.

15 "Indigenous Law Journal," University of Toronto Faculty of Law, accessed 8 August 2022, ilj.law.utoronto.ca/. The *Indigenous Law Journal* homepage is found here.

16 "June Callwood Program in Aboriginal Law," University of Toronto Faculty of Law, accessed 8 August 2022, perma.cc/H4JJ-MMVU. Information about the June Callwood Program in Aboriginal Law can be found here.

17 Kelly Gallagher-Mackay, *Succeeding Together: Schools, Child Welfare, and Uncertain Public Responsibility for Abused or Neglected* Children (Toronto: University of Toronto Press, 2017); Kelly Gallagher-Mackay and Nancy Steinhauer, *Pushing the Limits: How Schools Can Prepare Our Children Today for the Challenges of Tomorrow* (Toronto: Doubleday Canada, 2017). Professor Gallagher-Mackay now teaches at Wilfrid Laurier University. For a sample of her writing, see above references.

18 Kelly Gallagher-Mackay, "Affirmative Action and Aboriginal Government: The Case for Legal Education in Nunavut," *Canadian Journal of Law and Society* 14, no. 2 (1999): 21.

19 Shelly Wright, "The Akitsiraq Law School: A Unique Approach to Indigenous Legal Education," *Indigenous Law Bulletin* 5, no. 19 (2002): 14.

20 "Igloolik Elder Praised for Preserving Inuit Justice, Law," *CBC News*, 11 October 2007, perma.cc/93T5-APUR. This includes information about Lucien Ukaliannuk.

21 "Future Akitsiraq Law Grads Get First Look at Campus," *UVic News*, 25 January 2005, perma.cc/8RTH-8CMA.

22 Patricia A. Monture, "Women's Words: Power, Identity, and Indigenous Sovereignty," *Canadian Woman Studies* 26, nos. 3–4 (2008): 154; Patricia A. Monture, "Ka-Nin-Geh-Heh-Gah-E-Sa-Nonh-Yah-Gah," *Canadian Journal of Women and the Law* 2, no. 1 (1986): 159; Patricia Monture-Angus, *Thunder in My Soul: A Mohawk Woman Speaks* (Halifax: Fernwood, 1995); Patricia Monture-Angus, "Standing against Canadian Law: Naming Omissions of Race, Culture and Gender," *Yearbook of New Zealand Jurisprudence* 2 (1998): 7; Patricia Monture-Angus, "Women and Risk: Aboriginal Women, Colonialism, and Correctional Practice," *Canadian Woman Studies* 19, nos. 1–2 (1999): 24; Patricia A. Monture-Okanee, "The Violence We Women Do: A First Nations View," in *Challenging Times: The Women's Movement in Canada and the United States*, ed. Constance Backhouse and David H. Flaherty (Montreal: McGill-Queen's University Press, 1992), 193; Patricia A. Monture-Okanee, "The Roles and Responsibilities of Aboriginal Women: Reclaiming Justice," *Saskatchewan Law Review* 56, no. 1 (1992): 237. These are samples of Professor Monture's scholarship.

23 James (sákéj) Youngblood Henderson, "Empowering Treaty Federalism," *Saskatchewan Law Review* 58 (1994): 241; James Youngblood Henderson, *First Nations Jurisprudence and Aboriginal Rights: Defining the Just Society* (Saskatoon: Native Law Centre, 2006); James (sákéj) Youngblood Henderson, "Mikmaw Tenure in Atlantic Canada," *Dalhousie Law Journal* 18, no. 2 (1995): 196. These are samples of Professor Henderson's scholarship.

24 Harold Cardinal, *The Unjust Society: The Tragedy of Canada's Indians* (Edmonton: M.G. Hurtig, 1969); Harold Cardinal and Walter Hildebrandt, *Treaty Elders of Saskatchewan: Our Dream Is That Our Peoples Will One Day Be*

Clearly Recognized as Nations (Calgary: University of Calgary Press, 2000). These are samples of Harold Cardinal's work.

25 Ruth Thompson, "The University of Saskatchewan Native Law Centre," *Dalhousie Law Journal* 11, no. 2 (1988): 712; "Program of Legal Studies for Native People Gets New Name," University of Saskatchewan College of Law, 13 December 2017, perma.cc/BK78-JLGW; "University of Saskatchewan College of Law Offers 1L Indigenous Students Summer Classes," Indigenous Law Centre, 8 August 2022, perma.cc/S44K-5XVG. For historic information about the Program of Legal Studies for Native People, see Thompson article. Subsequent developments can be examined at "Program of Legal Studies for Native People Gets New Name." Spring and summer law courses for Indigenous students can be found at "University of Saskatchewan College of Law Offers 1L Indigenous Students Summer Classes."

26 John J. Borrows, "A Genealogy of Law: Inherent Sovereignty and First Nations Self-Government," *Osgoode Hall Law Journal* 30, no. 2 (1992): 291; John Borrows, "Negotiating Treaties and Land Claims: The Impact of Diversity within First Nations Property Interests," *Windsor Yearbook of Access to Justice* 12 (1992): 179. Publications drawn from my graduate work include those cited above.

27 John Borrows, *Recovering Canada: The Resurgence of Indigenous Law* (Toronto: University of Toronto Press, 2002); John Borrows, *Canada's Indigenous Constitution* (Toronto: University of Toronto Press, 2010).

28 Borrows, *Canada's Indigenous Constitution*, 228–36. The outline of the Indigenous law curriculum, which we now teach at UVic Law School, is detailed here.

29 Andrew Petter, *The Politics of the Charter: The Illusive Promise of Constitutional Rights* (Toronto: University of Toronto Press, 2010). Prior to serving as dean of UVic Law School, Andrew Petter had served in the Government of British Columbia as attorney general, minister of finance, minister of forests, minister of advanced education, minister of health, and minister of Aboriginal relations. His academic work is represented in *The Politics of the Charter*.

30 Avigail Eisenberg, Jeremy Webber, Glen Coulthard, and Andrée Boiselle, eds., *Recognition versus Self-Determination: Dilemmas of Emancipatory Politics* (Vancouver: UBC Press, 2014); Hester Lessard, Rebecca Johnson, and Jeremy Webber, eds., *Storied Communities: Narratives of Contact and Arrival in Constituting Political Community* (Vancouver: UBC Press, 2011); Jeremy Webber and Colin M. Macleod, eds., *Between Consenting Peoples: Political Community and the Meaning of Consent* (Vancouver: UBC Press, 2010); Hamar Foster, Heather Raven, and Jeremy Webber, eds., *Let Right Be Done: Aboriginal Title, the* Calder *Case, and the Future of Indigenous Rights* (Vancouver: UBC Press, 2007). Professor Webber has written about Indigenous legal issues.

31 James Tully, *Public Philosophy in a New Key* (New York: Cambridge University Press, 2008); James Tully, *Strange Multiplicity: Constitutionalism in an Age of*

Diversity (Cambridge: Cambridge University Press, 1995). Professor Tully was a special adviser to the Royal Commission on Aboriginal Peoples from 1991 to 1995. His publications also substantially engage with Indigenous issues.

32 "The First Nations Connection," *Canadian Lawyer*, 4 September 2007, perma.cc/PQ6S-QEFA. This is a discussion of Doug White's leadership as a student at UVic Law School.

33 Angela Cameron, Sari Graben, and Val Napoleon, eds., *Creating Indigenous Property: Power, Rights, and Relationships* (Toronto: University of Toronto Press, 2020); Hadley Friedland and Val Napoleon, "Gathering the Threads: Developing a Methodology for Researching and Rebuilding Indigenous Legal Traditions," *Lakehead Law Journal* 1, no. 1 (2015): 16; Val Napoleon and Hadley Friedland, "An Inside Job: Engaging with Indigenous Legal Traditions through Stories," *McGill Law Journal* 61, no. 4 (2016): 725. These are some recent publications by Val Napoleon.

34 "Indigenous Law Research Unit," ILRU, accessed 8 August 2022, perma.cc/S2FY-AUZ2. The Indigenous Law Research Unit's website can be found at ilru.ca.

35 Serena Dykstra, Zachary Donofrio, and Jasleen Johal, "Anishinaabe Law Camp," *Obiter Dicta*, 29 September 2014, perma.cc/B5VC-JNUC.

36 "Anishinaabe Law Class: Law as a Human Experience," *McGill Focus Law*, 1 November 2017, perma.cc/6BGK-3VYW.

37 "Indigenous Law Camp Imparts Valuable Lessons," *Western Law Faculty News*, 29 March 2017, perma.cc/UCT8-BAJX.

38 Hanna Askew, "Learning from the Land: Anishinaabe Law Camp at Walpole Island First Nation," *West Coast Environmental Law Blog*, 15 May 2016, perma.cc/P3ND-FQBM.

39 Meena Sundararaj, "A Student Writes: Indigenous Law in Context at Neyaashiinigmiing Cape Croker Reserve," *University of Toronto News*, 13 October 2017, perma.cc/4285-H2S3.

40 Borrows, *Law's Indigenous Ethics*, 149–75. This reference includes a discussion of land-based legal education.

41 Ryan Beaton, "Positivist and Pluralist Trends in Canadian Aboriginal Law: The Judicial Imagination and Performance of Sovereignty in Indigenous-State Relations" (PhD diss., University of Victoria, 2021).

42 Kinwa Kaponicin Bluesky, "Art as My Kabeshinan of Indigenous Peoples" (LLM thesis, University of Victoria, 2006).

43 Andrée Boisselle, "Law's Hidden Canvas: Teasing Out the Threads of Coast Salish Legal Sensibility" (PhD diss., University of Victoria, 2017).

44 Keith Cherry, "Practices of Pluralism: A Comparative Analysis of Trans-systemic Relationships in Europe and on Turtle Island" (PhD diss., University of Victoria, 2020).

45 Robert Clifford, "WSÁNEĆ Law and the Fuel Spill at Goldstream" (LLM thesis, University of Victoria, 2014).

46 Aimée Craft, "Breathing Life into the Stone Fort Treaty" (LLM thesis, University of Victoria, 2011).

47 Alan Hanna, "Dechen ts'edilhtan: Implementing Tsilhqot'in Law for Watershed Governance" (PhD diss., University of Victoria, 2020).

48 Carwyn Jones, "The Treaty of Waitangi Settlement Process in Māori Legal History" (PhD diss., University of Victoria, 2013).

49 Dawnis Kennedy, "Aboriginal Rights, Reconciliation and Respectful Relations" (LLM thesis, University of Victoria, 2009).

50 Darcy Lindberg, "Nêhiyaw Âskiy Wiyasiwêwina: Plains Cree Earth Law and Constitutional/Ecological Reconciliation" (PhD diss., University of Victoria, 2020).

51 Danika Billie Littlechild, "Transformation and Re-formation: First Nations and Water in Canada" (LLM thesis, University of Victoria, 2014).

52 Johnny Mack, "Thickening Totems and Thinning Imperialism" (LLM thesis, University of Victoria, 2009).

53 Aaron James Mills (Waabishki Ma'iingan), "Miinigowiziwin: All That Has Been Given for Living Well Together: One Vision of Anishinaabe Constitutionalism" (PhD diss., University of Victoria, 2019).

54 Morales, "*Snuw'uyulh.*"

55 Valerie Ruth Napoleon, "Ayook: Gitksan Legal Order, Law, and Legal Theory" (PhD diss., University of Victoria, 2009).

56 Joshua Nichols, "Reconciliation and the Foundations of Aboriginal Law in Canada" (PhD diss., University of Victoria, 2016).

57 Nicole Colleen O'Byrne, "Challenging the Liberal Order Framework: Natural Resources and Métis Policy in Alberta and Saskatchewan (1930–1948)" (PhD diss., University of Victoria, 2014).

58 Jacinta Arianna Ruru, "Settling Indigenous Place: Reconciling Legal Fictions in Governing Canada and Aotearoa New Zealand's National Parks" (PhD diss., University of Victoria, 2012).

59 Nancy Harriet Sandy, "Reviving Secwepemc Child Welfare Jurisdiction" (LLM thesis, University of Victoria, 2011).

60 Jennifer M. Sankey, "Globalization, Law and Indigenous Transnational Activism: The Possibilities and Limitations of Indigenous Advocacy at the WTO" (LLM thesis, University of Victoria, 2006).

61 Karen L. Sloan, "The Community Conundrum: Metis Critical Perspectives on the Application of R v Powley in British Columbia" (PhD diss., University of Victoria, 2016).

62 Tracey Lindberg, "Critical Indigenous Legal Theory Part 1: The Dialogue Within," *Canadian Journal of Women and the Law* 27, no. 2 (2015): 224; Brenda L. Gunn, "Remedies for Violations of Indigenous Peoples' Human Rights," *University of Toronto Law Journal* 69, no. S1 (2019): 150; Emily Snyder, "Challenges in Gendering Indigenous Legal Education: Insights from Professors Teaching about Indigenous Laws," *Canadian Journal of Law*

and Society 33, no. 1 (2019): 33. These are samples of this work at other Canadian law schools.

63 John Borrows, "Canada's Colonial Constitution," in *The Right Relationship: Reimagining the Implementation of Historical Treaties*, ed. John Borrows and Michael Coyle (Toronto: University of Toronto Press, 2017), 17.

64 Borrows, *Canada's Indigenous Constitution*, 12–22.

65 Borrows, 12–22.

66 John Borrows, "Sovereignty's Alchemy: An Analysis of *Delgamuukw v. British Columbia*," *Osgoode Hall Law Journal* 37, no. 3 (1999): 574–6.

67 Joe Sawchuk, "Social Conditions of Indigenous Peoples in Canada," *Canadian Encyclopedia*, last modified 28 August 2020, perma.cc/CRB6-VXJT.

68 Sawchuk, "Social Conditions of Indigenous Peoples in Canada."

69 Indigenous Services Canada, *Annual Report to Parliament 2020* (Ottawa: Indigenous Services Canada, 2020), 27.

70 Indigenous Services Canada, *Annual Report to Parliament 2020*, 17.

71 Indigenous Services Canada, 36.

72 Indigenous Services Canada, 51.

73 Naiomi Walqwan Metallic, "A Human Right to Self-Government over First Nations Child and Family Services and Beyond: Implications of the *Caring Society* Case," *Journal of Law and Social Policy* 28 (2018): 4; "It's Time for Fair Funding for First Nations Schools," #EndtheGap, accessed 2 September 2022, perma.cc/75CL-595L; Mohammad Hajizadeh, Min Hu, Amy Bombay, and Yukiko Asada, "Socioeconomic Inequalities in Health among Indigenous Peoples Living Off-Reserve in Canada: Trends and Determinants," *Health Policy* 122 (2018): 854.

74 Borrows, *Recovering Canada*, 111–37.

75 Gordon Christie, *Canadian Law and Indigenous Self-Determination: A Naturalist Analysis* (Toronto: University of Toronto Press, 2019). This includes a discussion of the problematic nature of Canadian law in relation to Indigenous peoples.

76 Borrows, *Law's Indigenous Ethics*, 88–113.

77 R v Sparrow, [1990] 1 SCR 1075 (Can.), 1109. In *R v Sparrow*, the Supreme Court found that section 35 demanded that the government justify "any ... regulation that infringes upon or denies aboriginal rights." It said that "such scrutiny is in keeping with the liberal interpretive principle enunciated in Nowegijick ... and the concept of holding the Crown to a high standard of honourable dealing with respect to the aboriginal peoples of Canada."

78 Sanderson, "Residue of *Imperium*," 331–2. However, there are arguments that Indigenous governance has been implicitly recognized by the courts.

79 Delgamuukw v British Columbia, [1997] 3 SCR 1010 (Can.), 1124.

80 Robert YELḰÁTⱩE Clifford, "W̱SÁNEĆ ('The Emerging People'): Stories and the Re-emergence of W̱SÁNEĆ Law," in *Renewing Relationships:*

Indigenous Peoples and Canada, ed. Karen Drake and Brenda L. Gunn (Saskatoon: Wiyasiwewin Mikiwahp Native Law Centre, 2019), 108–13.

81 Sarah Morales, "*Stl'ul Nup*: Legal Landscapes of the Hul'qumi'num Mustimuhw," *Windsor Yearbook of Access to Justice* 33, no. 1 (2016): 103; Nancy Sandy, "*Stsqey'ulécw Re St'exelcemc (St'exelemc* Laws from the Land)," *Windsor Yearbook of Access to Justice* 33, no. 1 (2016): 187. For discussions of Indigenous normativity in relation to broader views about life's purposes, which discuss who we are and where we are going, see article by Sarah Morales. For a similarly wide view, see Nancy Sandy's article.

82 Sara Florence Davidson and Robert Davidson, *Potlatch as Pedagogy: Learning through Ceremony* (Winnipeg: Portage and Main Press, 2018), 67–75; Gwi'molas Ryan Silas Douglas Nicolson, "'Playing the Hand You're Dealt': An Analysis of Musg̱amakw Dzawada'enux̱w Traditional Governance and Its Resurgence" (MA thesis, University of Victoria, 2019).

83 Corbin Greening, Lauren Mar, Ruben Tillman, and Calvin Sandborn, *The Case for a Guardian Network Initiative* (Victoria: Environmental Law Centre Society, 2020), perma.cc/EEA6-HDMP.

84 Alex Geddes, "Indigenous Constitutionalism beyond Section 35 and Section 91(24): The Significance of First Nations Constitutions in Canadian Law," *Lakehead Law Journal* 3, no. 1 (2019): 1; Mills, *Miinigowiziwin*, 38–9. Gabrielle Appleby, Vanessa MacDonnell, and Eddie Synot, "The Pervasive Constitution: The Constitution Outside of the Courts," *Federal Law Review* 48, no. 4 (2020): 437.

85 Chantelle Richmond, Heather Castleden, and Chelsea Gabel, "Practicing Self-Determination to Protect Indigenous Health in COVID-19: Lessons for This Pandemic and Similar Futures," in *COVID-19 and Similar Futures: Pandemic Geographies*, ed. Gavin J. Andrews, Valorie A. Crooks, Jamie R. Pearce, and Jane P. Messina (Cham, Switzerland: Springer, 2021), 307; Danielle Hiraldo, Kyra James, and Stephanie Russo Carroll, "Case Report: Indigenous Sovereignty in a Pandemic: Tribal Codes in the United States as Preparedness," *Frontiers in Sociology* 6 (2021): 1.

86 Aaron Mills, "The Lifeworlds of Law: On Revitalizing Indigenous Legal Orders Today," *McGill Law Journal* 61, no. 4 (2016): 847. For example, Aaron Mills, who now works at McGill's Faculty of Law, proposed that law schools should teach that all law is storied, that Canadian constitutional law is a species of liberal constitutionalism, and that students should take a prerequisite on an Indigenous people's constitutional order before enrolling in a course on their law.

87 Karen Drake, "Finding a Path to Reconciliation: Mandatory Indigenous Law, Anishinaabe Pedagogy, and Academic Freedom," *Canadian Bar Review* 95 (2017): 9.

88 *Truth and Reconciliation Commission of Canada: Calls to Action* (Winnipeg: Truth and Reconciliation Commission, 2015), 3. Kathleen Mahoney,

"Indigenous Legal Principles: A Reparation Path for Canada's Cultural Genocide," *American Review of Canadian Studies* 49, no. 2 (2019): 207. See Kathleen Mahoney article for commentary about the importance of Indigenous law to the truth and reconciliation process.

89 Aaron Mills, Karen Drake, and Tanya Muthusamipillai, "An Anishinaabe Constitutional Order," in *Reconciliation in Canadian Courts: A Guide for Judges to Aboriginal and Indigenous Law, Context and Practice*, ed. Justice Patrick Smith (Ottawa: National Judicial Institute, 2017), 260. Another way of teaching Anishinaabe law is discussed here.

90 Leaelle N. Derynck, "An Anishinaabe Tradition: Anishinaabe Constitutions in Ontario" (LLM thesis, University of Western Ontario, 2020).

91 Larry Nesper, "Negotiating Jurisprudence in Tribal Court and the Emergence of a Tribal State: The Lac du Flambeau Ojibwe," *Current Anthropology* 48, no 5. (2007): 675; Matthew L.M. Fletcher, "*Laughing Whitefish*: A Tale of Justice and Anishinaabe Custom," *Michigan State University College of Law Legal Studies Research Paper Series* no. 06–16 (2008): 10–11. Matthew L.M. Fletcher, *The Eagle Returns: The Legal History of the Grand Traverse Band of Ottawa and Chippewa Indians* (East Lansing: Michigan State University Press, 2012), 2–33. For a discussion of Anishinaabe law in the United States, see articles by Larry Nespar and Matthew Fletcher. The resurgence of Anishinaabe responsibilities in one community is described in Matthew Fletcher's book.

92 Malcolm Lavoie and Moira Lavoie, "Indigenous Institutions and the Rule of Indigenous Law," *Supreme Court Rule of Law* 101 (2021): 325. See generally this article for context surrounding this dispute.

93 Roxanne Egan-Elliott and Louise Dickson, "Protesters Block Legislature Entrances, MLAs Have to Squeeze By to Hear Throne Speech," *Times Colonist*, 11 February 2020, perma.cc/DFR9–6RWP; "The Spatial Politics of Energy Conflicts: How Competing Constructions of Scale Shape Pipeline and Shale Gas Struggles in Canada," *Energy Research and Social Science* 77 (2021): 1. For a discussion of the site-specific nature and spatialization of Indigenous politics, see Carol Hunsberger and Rasmus Kløcker Larsen article.

94 Richard Watts, "UVic Students Walkout in Solidarity with Wet'suwet'en Chiefs over Pipeline," *Times Colonist*, 10 January 2020, perma.cc/EKK6-UXTT; Canadian Press, "Horgan Says Pipeline Protests at B.C. Legislature 'Counterproductive,'" *CTV News*, 4 March 2020, perma.cc/6MDU-ZN53; Canadian Press, "Protesters Pack Up Camp at B.C. Legislature after Five Arrests Wednesday Night," *CKPGToday*, 7 March 2020, perma.cc/3GJJ-63G4; Kathleen Mahoney, "Indigenous Laws and Human Rights Uprisings," *Journal of Diplomacy and International Relations* 21, no. 2 (2020): 108. For a discussion of Indigenous law and energy conflicts, see Kathleen Mahoney.

95 Mentor L. Williams, ed., *Schoolcraft's Indian Legends* (East Lansing: Michigan State University Press, 1956), 106–8.

96 John Borrows, *Freedom and Indigenous Constitutionalism* (Toronto: University of Toronto Press, 2016), 209–14.

97 Matthew L.M. Fletcher, "Indian Courts and Fundamental Fairness: Indian Courts and the Future Revisited," *University of Colorado Law Review* 84, no. 1 (2013): 91n128; Champagne v People (of the Little River Band of Ottawa Indians), 2007, Little River Band of Ottawa Indians Tribal Court of Appeal Case No 06–178-AP; Matthew Fletcher, *American Indian Tribal Law* (New York: Aspen Publishers, 2011), 405–12. See Anishinaabe court cases dealing with section 7–like rights in the *Charter* in Fletcher, "Indian Courts and Fundamental Fairness" which reads,

Spurr, No. 12–005APP, at 7 (citing Crampton v. Election Bd., 8 Am. Tribal Law 295, 296 (Little River Band of Ottawa Indians Tribal Ct. May 8, 2009); Bailey v. Grand Traverse Band Election Bd., No. 2008–1031 -CV-CV, 2008 WL 6196206, at *9, 11 (Grand Traverse Band of Ottawa and Chippewa Indians Tribal Judiciary, Aug. 8, 2008) (en banc); Deckrow v. Little Traverse Bay Bands of Odawa Indians, No. C-006–0398, 1999 WL 35000425, at *2 (Little Traverse Bay Bands of Odawa Indians Tribal Ct. Sept. 30, 1999)).

98 Carter v Canada (Attorney General), 2015 SCC 5; R v Morgentaler, [1988] 1 SCR 30; Rodriguez v British Columbia (Attorney General), [1993] 3 SCR 519; Canada (Attorney General) v Bedford, 2013 SCC 72. In particular, we consider *Carter v Canada* and *R v Morgentaler* in the context of *Rodriguez v British Columbia (Attorney General)* and *Canada (Attorney General) v Bedford*.

99 Reverend Frederic Baraga, *A Dictionary of the Otchipwe Language* (Cincinnati, OH: Jos A Hermann, 1853), s.v. "dibénindisowin"; *The Ojibwe People's Dictionary*, s.v. "dibenindizowin," accessed 4 September 2022, perma.cc/K63G-L45X. Anishinaabemowin is the original language of the northern and western Great Lakes of North America. Citations to *dibenindisowin* can be found by Reverend Frederic Baraga, stating that the word means "liberty, freedom, independence." The second reference gives the meaning "s/he is independent, is h/ own master" for the word *dibenindizowin*).

100 Basil H. Johnston, *Anishinaubae Thesaurus* (East Lansing: Michigan State University Press, 2007), s.v. "dae'b'ingaewin," 73; Borrows, *Law's Indigenous Ethics*, 53–6. In Basil Johnston's thesaurus, *dae'b'ingaewin* is "to tell what one knows according to his/her perception and according to one's fluency; to have the highest degree of accuracy; to be right, correct; to have truth."

101 Nicolas Perrot, *The Indian Tribes of the Upper Mississippi Valley and Region of the Great Lakes*, ed. and trans. Emma Helen Blair, vol. 1 (Cleveland, OH: Arthur H. Clark, 1911), 35–6; Pierre François Xavier de Charlevoix, *Journal of a Voyage to North America*, ed. Louise Phelps Kellogg, vol. 1 (Chicago: Caxton Club, 1923), 155–6. Borrows, *Canada's Indigenous Constitution*, 331–2. Muskrat is a hero in Anishinaabe creation stories

102 Borrows, *Law's Indigenous Ethics*, 114.

103 Borrows, 24–49.

104 Wendy Makoons Geniusz, "Manidoons, Manidoosh: Bugs in Ojibwe Culture," in *Papers of the Forty-Fourth Algonquian Conference*, ed. Monica Macaulay, Margaret Noodin, and J. Randolph (Albany: SUNY Press, 2016), 99.

105 Borrows, *Law's Indigenous Ethics*, 176–8. For information about bears in Anishinaabe law, see reference.

106 Basil Johnston, *The Gift of the Stars (Anungook gauh meenikooying)* (Cape Croker First Nation, ON: Kegedonce Press, 2010); Annette Sharon Lee, William Peter Wilson, and Carl Gawboy, *Ojibwe Sky Star Map – Constellation Guidebook: An Introduction to Ojibwe Star Knowledge* (Minneapolis: Native Skywatchers, 2014).

107 Borrows, *Law's Indigenous Ethics*, 150.

108 Lindsay Keegitah Borrows, *Otter's Journey through Indigenous Language and Law* (Vancouver: UBC Press, 2018).

109 John Borrows (Kegedonce), *Drawing Out Law: A Spirit's Guide* (Toronto: University of Toronto Press, 2010), 14–16.

110 Borrows, *Law's Indigenous Ethics*, 162.

111 Borrows, 162; Edward Benton-Banai, *The Mishomis Book: The Voice of the Ojibway* (St. Paul, MN: Red School House, 1988).

112 D'Arcy Rheault, *Anishinaabe Mino-Bimaadiziwin: The Way of a Good Life* (Peterborough, ON: Debwewin Press, 1999); Winona LaDuke, "Minobimaatisiiwin: The Good Life," *Cultural Survival Quarterly* 16, no. 4 (1992): 69.

113 Borrows, *Law's Indigenous Ethics*, 92.

114 Borrows, 239–40.

115 Joshua Ben David Nichols, *A Reconciliation without Recollection? An Investigation of the Foundations of Aboriginal Law in Canada* (Toronto: University of Toronto Press, 2019); William Nikolakis, Stephen Cornell, and Harry W. Nelson, eds., *Reclaiming Indigenous Governance: Reflections and Insights from Australia, Canada, New Zealand, and the United States* (Tucson: University of Arizona Press, 2019). Finding space for Indigenous law and governance is not easy when Canadian law continues to dominate. As students identify these and other ambiguities, this helps us search for clarity. Incongruity can prompt a search for clarification. Acknowledging uncertainty can be an invitation to study. Experimentation must contemplate the possibility of failure, and I have had my share of disappointments working to enhance Indigenous law. The biggest problem is the work's vast scale. For instance, legislatures and courts do not know enough about Indigenous law, and many Indigenous communities struggle to escape colonialism's grasp.

PART ONE

PRINCIPLES

(In)Voluntarily Enfranchised: Bill C-3 and the Need for Strengthening Kinship Laws in Treaty 4

DANETTE JUBINVILLE

I wrote this essay in 2015, when I was newly pregnant with my daughter. Since then, I have connected with several families who reached out to me after finding my essay on RAVEN's website while doing research into their own status disputes. Some expressed that my essay was the only literature they had been able to find on the issue of voluntary enfranchisement. In August of 2020, a lawyer named David Schulze emailed me to share that he had read my essay while working on a status case for his client, Karl Hele, whose mother had enfranchised by application. He wanted me to know that they had won their case just a few days prior, and that I should now be eligible for status.

The argument put forward in Hele c Canada *centred around the issue of gender; Schulze and company argued that* Indian Act *enfranchisement policies were explicitly gendered and that the section that dealt with voluntary enfranchisements was worded in a way that only applied to Indian men. The judge agreed that* Indian Act *administrators had erred in processing enfranchisement applications made by women. This decision set a precedent to amend the registry records for all women who had "voluntarily" enfranchised, including my grandmother. As a result of this decision, I am now eligible for Indian status.*

In September of 2021, we are still waiting for the registrar to amend my grandmother's record as well as the Indian Act *subcategory under which my father and his siblings are registered, so that my and my cousins' applications can be processed. While I sit in frustration that we are still waiting for this error and act of discrimination to be resolved, I am humbled that my essay played some small part in this precedent-setting case, and grateful to everyone involved with the RAVEN award who made it possible. I would also like to express my gratitude here to the Pasqua First Nation Chief and Council for recognizing my grandmother's story as an issue of discrimination and for working through their own channels to try and make it possible for us to be eligible for status and band membership prior to the* Hele *decision.*

Figure 3.1. Danette's daughter Keestin at Asham's Beach, Pasqua First Nation, 2019.

Source: Courtesy of Danette Jubinville.

Introduction

At the present moment, there are thousands of people across Canada imagining, articulating, questioning, living, and strengthening Indigenous nationhood. Central to this work is determining who belongs in Indigenous nations.[1] As a "non-status Indian" with Cree and Anishinaabe blood, this question is not only at the forefront of my research, but also my identity. The *Indian Act* fails to address this question in a way that reflects Indigenous peoples' own understandings of citizenship and belonging, and so I have to look for solutions outside it.

In this essay, I explore the need for strengthening Cree and Anishinaabe kinship laws and codes of belonging in Treaty 4 by examining the *Indian Act* concept of voluntary enfranchisement. I look at the gendered aspects of voluntary enfranchisement, and how, as in my family's case, this kind of legislation disempowers women. I critique Bill C-3, the *Gender Equity in Indian Registration Act,* for failing to address the issue of voluntary enfranchisement. Finally, I argue that in order to maintain the livelihood of our Treaty 4 nations, it is necessary to reject *Indian Act*–defined codes of membership and reinstate our own citizenship laws. Furthermore, this essay asserts that reinvigorating traditional citizenship laws requires the fulfilment of Treaty 4 rights on the terms that were agreed upon, separate from the *Indian Act.*

Personal Implications

I am interested in this topic because it is a legal matter that has had a tremendous impact on my family and myself. Enfranchisement is the reason why I do not have Indian status or band membership in my father's community, the Pasqua First Nation. As a result, I am not a beneficiary of Treaty 4, even though my relatives were promised that the treaty would take care of their children "as long as the sun shines and water flows,"[2] in exchange for sharing our homelands with the European settlers.

My grandmother belongs to the Cyr family from Pasqua, a Cree and Saulteaux community outside of Fort Qu'appelle, Saskatchewan. The community is named after Chief Pasqua, our last traditionally appointed chief and signatory of Treaty 4. My grandmother was born on the reserve into the hands of her grandmother, Isabel Bear, in 1933. As a child, she was taken to the Lebret Indian Residential School, where she was forced to stay for ten years. My grandmother was a bright student, and the nuns encouraged her to train as a nurse technician in Manitoba. After graduation, she followed their advice, and while training she fell in love with my grandfather Léandre Jubinville, a French Canadian. He asked for her hand in marriage, and they agreed to wait two years, until the spring of 1955, after they both finished their post-secondary training.

Six months before her wedding date, my twenty-one-year-old grandmother returned home to Pasqua. While she was home, the Indian agent paid her a visit to remind her that her upcoming marriage to a non-Native man would result in the automatic loss of her Indian status, in accordance with the *Indian Act* at the time. He gave her paperwork, explaining that in exchange for her Indian status, she would receive a one-time payment. She knew that she had no choice about losing her status, and the payment would help pay for the wedding. Although my grandmother doesn't believe she received any payment in the end, the Indian registrar alleges that on 3 November 1955, my grandmother voluntarily gave up her Indian status, or, in legal terms, she "enfranchised by application."

My grandmother's Indian status, as well as my father's, was restored in 1985 under Bill C-31, the amendment that struck enfranchisement rules from the *Indian Act*.[3] According to the Bill C-31 amendments, my grandmother was registered under the provisions of section 6(1)(d) of the *Indian Act*, and my father was registered under section 6(2). Registration under section 6(2) means that one can only pass on Indian status to one's children if the other parent is also a status Indian. Since my mother is not Indigenous, I am not entitled to be registered under the act.

My family's situation raises many questions about justice. Was my grandmother's supposedly "voluntary" decision to enfranchise legally binding? How can the logic of enfranchisement be applied to my generation, even though it no longer exists within the *Indian Act*? And why do I even need Indian status to receive treaty benefits?

Indian Act Enfranchisement

The term "enfranchisement" was first introduced in Canadian Indian policy in 1857, via the *Gradual Civilization Act*.[4] Enfranchisement "meant that an Indian was no longer an Indian in law, had become civilized, and was entitled to all the rights and responsibilities of other Canadian citizens."[5] In order to enfranchise, an Indian man had to prove that he was "reasonably well educated, free of debt, and of good moral character as determined by a commission of non-Indian examiners."[6] The 1857 enfranchisement laws targeted males and emphasized the idea of "civilization" through the ownership of private property. These policies represented the beginning of a trend towards "better management" and control over Indigenous peoples in settler-state law, and built on the reports and recommendations of a Commission of Inquiry established in 1841 to "investigate the condition of the Indians."[7] The 1996 Royal Commission on Aboriginal Peoples characterized enfranchisement as "a euphemism for one of the most oppressive policies adopted by the Canadian government in its history of dealings with aboriginal peoples."[8]

In 1869, two years after Canadian Confederation, the *Gradual Enfranchisement Act* was passed. This act penalized Indian women who married non-Indian men by automatically enfranchising them as Canadians.[9] In the case of non-Indian women marrying Indian men, the opposite rule applied. Non-Indian women who married Indian men gained Indian status for themselves and their children. As far as settler-state policy was concerned, Indian women were the property of men.

More elaborate forms of control over the definition of "Indian" were introduced with the *Indian Act* of 1876. Other Indians who were automatically enfranchised included "professionals," meaning lawyers, doctors, ministers, or anyone with a university degree.[10] Prior to 1956, it was not possible to simultaneously be an Indian and a Canadian citizen in Canadian law,[11] and other reasons for voluntary enfranchisement may have included the desire to enlist in the army or vote in federal elections, which Indians could not do until 1960.[12] Between 1918 and 1922, the Canadian government granted itself the right to enfranchise any Indian who met the requirements for voluntary enfranchisement, but this law was overturned due to opposition, and application was again necessary.[13]

In 1951, the membership and enfranchisement sections of the *Indian Act* were changed significantly to the detriment of Indian women.[14] Prior to 1951, Indian women who lost their status by marrying non-Indians could still obtain some treaty and band rights with a "Red Ticket," though they and their children lost the right to live on reserve.[15] After 1951, Indian women who "married out" lost their status, band, and treaty rights. An enfranchised woman would have thirty days to dispose of any property she held on the reserve, as she could no longer live there.[16] Applications to enfranchise could still be made under the 1951 legislation, pending proof of self-sufficiency and the consent of the band.[17]

Between 1856 and 1985, very few Indians enfranchised voluntarily, in comparison to involuntary enfranchisements. Only one Indian enfranchised between the passing of the *Gradual Civilization Act* of 1856 and the *Indian Act* of 1876.[18] Though relatively small, the number of Indians who enfranchised "voluntarily" after 1876 is likely misleading, as many of these enfranchisements may have happened prior to marrying out, as in the case of my grandmother. As I will discuss later, this decision was not voluntary at all, given that there was no other option.

Enfranchisement policies continued until 1985, when the Canadian government passed Bill C-31 in response to Indigenous women's activism and litigation efforts to end gender discrimination within the *Indian Act*, which also generated international pressure.[19] Additionally, the Canadian government was trying to repatriate the constitution, and the proposed *Charter of Rights and Freedoms* required that laws apply equally towards men and women.[20] Bill C-31 ended involuntary enfranchisement through marriage, as well as enfranchisement by application. As a result, my grandmother got her Indian status and band membership back, and my father's status and membership were recognized as well. However, the "second-generation cut-off" rule meant that I did not qualify for status.

Bill C-31 did not meaningfully address gender discrimination, because the "second-generation cut-off" rule only applied to the grandchildren of women who got their status back through Bill C-31, whose parents were reinstated under the new and lesser category of section 6(2). This is still gender discrimination, because if it was my grandfather who was

Indian instead of my grandmother, my father would have been a section 6(1) Indian, with the ability to pass on his status (and, if my grandfather was Indian, no one in our lineage would have lost their status and band membership to begin with, because he never would have faced the possibility of losing his status through marriage to a non-Indian). Indigenous women continued to fight this injustice, and the court case *McIvor v Canada* led to the passing of Bill C-3, the *Gender Equity in Indian Registration Act*, in 2011. Under Bill C-3, the second-generation cut-off rule was amended to include another generation (but then cut off).

Bill C-3, the *Gender Equity in Indian Registration Act*

Specifically, Bill C-3 grants status to those who meet the following criteria: (1) the "individual's grandmother lost her Indian status as a result of marrying a non-Indian man"; (2) "one parent is registered, or entitled to be registered, under sub-section 6(2) of the *Indian Act*"; and (3) the "individual or one of the individual's siblings was born on or after September 4, 1951."[21] Letters addressed to me from Aboriginal Affairs and Northern Development Canada (AANDC) reference the criteria set out by Bill C-3, and state "[Your paternal grandmother …] lost her entitlement to Indian status under the provisions of a former *Indian Act* when she enfranchised by application on 1955/11/03."[22] The letter goes on to explain that "There is no provision in the current *Indian Act* to allow for the registration of a person when one of the parents is registered under Section 6(2) of the *Indian Act* and the other parent is not entitled to be registered under the *Indian Act*."[23]

Despite the fact that my grandmother did, in every practical sense, lose her entitlement to registration as a status Indian as a result of marrying a non-Indian, AANDC does not interpret her application to enfranchise this way. Because AANDC narrowly interprets the criteria of Bill C-3 to only include those women who lost their status automatically through marriage, those grandchildren of women who lost their status by application, prior to marrying out, are not eligible for status under this bill.

While there are no published critiques about how voluntary enfranchisement is inextricably tied to involuntary enfranchisement, or how Bill C-3 fails to address the issue of "voluntary" enfranchisement, the bill has been widely criticized. Bill C-3 as it is written does not truly address the issue of gender discrimination; it just puts off the problem for another generation.[24] In addition, the bill has many shortcomings, including that it replicates the same logic it claims to overcome, it adds confusing wording to an already confusing piece of legislature, it "does not provide adequate protections" for new registrants with regards to band membership, and it is still in conflict with the *Charter of Rights and*

Freedoms.[25] In spite of the fact that enfranchisement is no longer possible, this unconstitutional logic is still applied to thousands of Indigenous people across Canada, including myself.

Indigenous women in particular have fought the gender-discriminatory nature of the *Indian Act* in domestic and international settings, and still continue to deal with the psychological and material damage done by this legislation. Most gendered analyses of the *Indian Act* focus on involuntary enfranchisement through marriage, marital property rights, and the "double mother clause," but almost no sources look at the gendered dimension of enfranchisement by application. My grandmother's story shows that it is likely many of those women who supposedly enfranchised voluntarily actually had no choice in the matter, as they were about to lose their status involuntarily. While Bill C-3 granted status to the grandchildren of women who were involuntarily enfranchised through marriage to a non-Indian, no explicit rules apply for the grandchildren of women and men who enfranchised by application, and therefore our status applications continue to be denied. Voluntary enfranchisement, then, is another one of the many ways that the *Indian Act* has targeted the removal of Indigenous women from Indigenous lands, and through which the federal government excuses itself from its fiduciary and legal obligations to Indigenous peoples.

"Indian," the *Indian Act*, and Treaty

By creating the tightly controlled legal category "Indian," Canada found a way to slowly but surely legislate itself out of its responsibilities to the First Peoples of this land. After almost one hundred and fifty years of implementation, government control over Indian identity is still uncomfortable, but in many ways normalized. It is important to examine what business Canada had in determining who is and is not an "Indian," as well as when, how, and why this came to be accepted by Indigenous peoples.

Building on the *Gradual Enfranchisement Act* of 1869 and the *Gradual Civilization Act* of 1856, the *Indian Act* "defined Indigenous peoples as 'Indians,' with criteria that specifically excluded Indigenous women and their children, as well as the female children of Indian men."[26] The *Indian Act* came into effect in Canadian law in 1876, two years after Treaty 4 was signed in Fort Qu'appelle, Saskatchewan, and shortly after the making of Treaty 6. Saskatchewan treaty historians Ray, Miller, and Tough note that

> Parliament codified legislation affecting First Nations in the Indian Act of 1876, the same year in which Plains Nations won major advances in the Treaty of Fort Carlton and Fort Pitt, Treaty 6. The latter document embodied a relationship between nations that contemplated a future based on dialogue

and accommodation. The former, the Indian Act, treat First Nations throughout Canada as legal minors and approached them as a problem to be administered.[27]

In the historical records of the Treaty 4 and 6 negotiations, no reference was made to "status Indians," enfranchisement, or the *Indian Act*, and there is no mention of these terms in the written treaty agreements either.[28]

Although there are no documented conversations about the *Indian Act* during treaty making, numerous records show that the colonial government clarified many times who was bound to the treaty. Promises similar to this appear several times in the shorthand reports of the proceedings submitted by the head treaty negotiator, Lieutenant Governor Alexander Morris:

> I will pass away and you will pass away. I will go after my fathers have gone and you also, but after me and after you will come our children. The Queen cares for you and for your children, and she cares for the children that are yet to be born. She would like to take you by the hand and do as I did for her at the Lake of the Woods last year. We promised them and we are ready to promise now, to give five dollars to every man, woman and child, and long as the sun shines and water flows.[29]

There are no qualifiers here, like "only status Indians" or "only Indians as defined by the Indian Act" will benefit from the treaty. The colonial government made it very clear that all the descendants of treaty communities are bound to the treaty.

While the subject of Métis people came up during both Treaty 4 and 6 talks, with the Indian chiefs asking for the Métis to be included and the colonial government promising to deal with them separately, it seems that these references to "Half-breeds" are specific to organized Métis communities, and not mixed-blood people more generally. The Treaty 6 record states

> A request was then made that the treaty should include the Half-breeds, to which the Governor replied: "I have explained to the other Indians that the Commissioners did not come to the Half-Breeds: there were however a certain class of Indian Half-breeds who had always lived in the camp with the Indians *and were in fact Indians*, would be recognized, but no others."[30]

At this time, Indian and Métis were not distinguished from one another based on race or blood quantum alone.[31] Rather, arbitrary lines were drawn based on several factors, including "biology, culture, lifestyle, moral comportment, and the ability to support oneself and one's family."[32] During treaty making, clear distinctions between Indian and Métis were never made, nor was access to treaty rights made contingent upon Indian status. Many people who considered themselves "Indian" took half-breed scrip, and many people who considered themselves "Métis" took treaty. Shortly afterwards, however, the legal-administrative categories of "Indian" and "half-breed" became more tightly controlled in order to extinguish claims to rights under those categories.[33]

The shift in attitudes that the *Indian Act* represented was served alongside the "collapse in of the Plains buffalo economy," and the "harsh atmosphere that prevailed after the Northwest Rebellion."[34] At the outset, implementation of the *Indian Act* on the ground must have been tentative at best, given that Indigenous peoples clearly saw themselves as sovereign at treaty negotiations, and that Canadian sovereignty was still "pretentious and largely fictitious."[35] In the years following Treaty 4, however, starvation on the plains, disease, tightening measures of enforcement, and the creation of the reserve system were some of the reasons why the *Indian Act* came be accepted as law.

Today, benefits associated with treaty negotiations, such as those related to education, health, hunting, and fishing, are administered through the *Indian Act*. Although it may not be the government's stated policy to administer treaty rights through the act, without Indian status, there is no alternative way to access treaty rights. The Canadian government has a vested interest in conflating treaty rights with *Indian Act* benefits. By controlling who is and who is not "Indian," Canada will eventually free itself of its treaty obligations. If there are no status Indians, then there are no annuities to be paid, or lands to reserve. However, the *Indian Act* is completely separate from treaty, and undermines not only the original spirit and intent of treaty making, but also the terms on which the treaties were signed.

It is imperative, then, that we understand what our treaty rights and responsibilities are, as well as who is bound to them. In Treaty 4, the treaties were signed with the understanding that we would govern ourselves, free from the colonial government's meddling. Therefore, we retain the right to determine who we are, who we belong to, and who is bound in responsibility to the treaty. Our own laws of belonging and membership can help us in this regard.

Kinship Laws in Treaty 4: Beyond the *Indian Act*

When thinking about who belongs in our Treaty 4 nations, it is necessary to understand the laws that informed our ancestors' negotiations. The first question we might ask is for whom were our ancestors negotiating?

Our chiefs signed the treaties on behalf of the families in their communities, and kinship laws formed the basis of belonging within those families. The difference between Indigenous and colonial views of citizenship within Indigenous nations was a point of contention; in the Treaty 4 records, leaders expressed their dissatisfaction with the government's exclusion of the Métis, stating that in their minds, the Métis, Crees, Saulteaux (Anishinaabe), and Stonies were one:

> The Gambler: Now when you have come here, you see sitting out there a mixture of Half-breeds, Crees, Saulteaux and Stonies, all are one, and you were slow in taking the hand of a Half-breed. All these things are many things that are in my way.[36]

In the pre-treaty period, identity in Treaty 4 was fluid and diverse, with Cree, Anishinaabe, Assiniboine, and Métis people living closely together, while still maintaining distinct stories and cultural practices.[37] Colonized thinking about identity and belonging does not reflect the diversity of culture, blood, and language that exists in Treaty 4.[38] Innes argues that kinship was the "central unifying factor" that allowed Indigenous peoples of the northern plains to "integrate others into their bands" and live relatively free of conflict.[39]

In Treaty 4, kinship laws formed the basis of nationhood, because they allowed citizens to intermarry outsiders, adopt enemies, spread across vast distances, incorporate a wide range of outsiders into families, and still remain unified.[40] Unlike the *Indian Act* system, kinship does not present two stark choices: to either assimilate into the dominant culture or be excluded from it. On the contrary, the flexibility of kinship laws, which are based on mutual responsibilities, allowed our people to survive.[41]

Today, Indigenous livelihood is compromised by a number of complex and overlapping factors, including oppressive *Indian Act* policies, consistent underfunding for Indigenous communities and populations, systemic racism, and the lasting effects of centuries of colonial violence. Yet, the queen's representatives promised the Treaty 4 leaders that their families would maintain a high quality of life, in perpetuity, as a result of those negotiations. In reality, the Canadian government has failed to implement the numbered treaties in a way that is true to what was said during treaty making. As a result, Indigenous peoples face the lowest quality of life of any population in Canada. While benefits associated with the treaties are administered haphazardly through the *Indian Act*, many Indigenous peoples cannot access those benefits, because of *Indian Act*–defined codes of membership and identity. It is important for Indigenous leadership to remember that the *Indian Act* did not exist at the time of treaty, and that enfranchisement policies, which continue today, and which discriminate against women, are one of the Canadian government's main strategies for escaping its treaty responsibilities. Planning, building, and fighting for healthy Indigenous communities in Treaty 4 will require strategic thinking towards creating the sense of belonging and unity that existed prior to 1876. To that end, we have to strengthen our ancestral kinship laws, while demanding that Canada fulfil its treaty and fiduciary obligations to all Indigenous peoples. We need our entire families to be resurgent nations, not just those who are status Indians.

Conclusion

I began this essay with the reason why I do not have Indian status and then examined the concept of "voluntary" enfranchisement in the *Indian Act*. I found that voluntary enfranchisement was not so voluntary after all, but was instead another way that the *Indian Act* brought about the removal of Indigenous women from Indigenous lands. I added to critiques of Bill C-3, the *Gender Equity in Indian Registration Act*, by pointing out that because the bill fails to address the discriminatory nature of voluntary enfranchisement, the logic of enfranchisement still exists. Next, I questioned the validity of the *Indian Act*, showing that Treaty 4 and the *Indian Act* are two completely separate things, the former not to be determined by the latter. Finally, I discussed kinship laws as the basis of belonging in a Treaty 4 community and found that these laws must be strengthened. At the same time, Canada must honour the treaties, according to the terms that were agreed upon.

I originally undertook this research to find a solution within Canadian law to gain Indian status, to which I believed my cousins and I were entitled. However, the research process led me to an alternate conclusion, which is that gaining Indian status is not the only important outcome. Indigenous communities have experienced immense change over the last two hundred years, yet Canadian legislation remains essentially the same. The intention to assimilate and eradicate the Indigenous population is still alive and well in state policy. Yet today, we have slightly more room to move than our ancestors did. Therefore, it is pertinent that we ask ourselves why we are allowing the *Indian Act* to shape our lives.

Our treaty rights and the *Indian Act* are two completely separate things. The colonial system will always act in a self-interested way, and the purpose of the *Indian Act* is to eventually enable Canada to legislate itself out of its treaty obligations. What would it look like if we reimagined our membership codes according to our own kinship laws and outside the confines of *Indian Act* logic? The *Indian Act* threatens our self-worth, our sense of belonging, and the unity within our communities. It undermines our spiritual and cultural ways of life by creating intense limitations on who we are, where we can go, and what we can become. The *Indian Act* was designed to undermine our livelihood, and it will do just that – it has done just that.

It may seem as if, with this essay, I am searching for a way to belong in Treaty 4. But in many ways, I know that I already do. My family accepts me as their own, I always have a place to stay in our territory, and when I mention my family name, I am instantly welcomed by people I have never met. In this way, we keep our Cree and Anishinaabe kinship laws alive. These are the laws that keep us full and accountable, and so these are the laws that we must invest in, for the future strength of our nations.

NOTES

1 Andrea Smith, *Conquest: Sexual Violence and American Indian Genocide* (Cambridge: South End Press, 2005), 97n54.
2 Alexander Morris, *The Treaties of Canada with the Indians of Manitoba and the North-West Territories: Including the Negotiations on Which They Were Based, and Other Information Relating Thereto* (Toronto: Belfords, Clarke and Co., 1880), 96.
3 Native Women's Association of Canada, *Guide to Bill C-31: An Explanation of the 1984 Amendments to the Indian Act* (Ottawa: Native Women's Association of Canada, 1986), 4.
4 Kathleen Jamieson, *Indian Women and the Law in Canada: Citizens Minus* (Ottawa: Minister of Supply and Services Canada, 1978), 27.
5 Jamieson, *Indian Women and the Law in Canada*, 13–14.
6 Royal Commission on Aboriginal Peoples, *Report of the Royal Commission on Aboriginal Peoples*, vol. 1, *Looking Forward, Looking Back* (Ottawa: Canada Communication Group Publishing, 1996), 249.
7 Jamieson, *Indian Women and the Law in Canada*, 22, 29.
8 Royal Commission on Aboriginal Peoples, *Report of the Royal Commission on Aboriginal Peoples*, 1:271.
9 Royal Commission on Aboriginal Peoples, 1:5.
10 Jamieson, *Indian Women and the Law in Canada*, 44.
11 Jamieson, 14.
12 Jamieson, 47.
13 Jamieson, 50.
14 Jamieson, 59.
15 Jamieson, 61.
16 Jamieson, 63.
17 Jamieson, 61.

18 Royal Commission on Aboriginal Peoples, *Report of the Royal Commission on Aboriginal Peoples*, 1:250.

19 *Indian Act*, RSC 1985, c I-5; Bill C-31, an *Act to Amend the Indian Act*, included various amendments to different sections, not just section 91.

20 "Enfranchisement," UBC First Nations Studies Program, accessed 7 April 2014, http://indigenousfoundations.arts.ubc.ca/home/government-policy /the-indian-act/enfranchisement.html.

21 AANDC, *Report to Parliament – Gender Equity in Indian Registration Act* (Ottawa: Aboriginal Affairs and Northern Development Canada, 2013).

22 T.M. Cloutier (deputy Indian registrar, AANDC), letter to the author, 27 January 2015.

23 Cloutier, letter to the author.

24 Pamela Palmater, "Bill C-3 Gender Equity in Indian Registration Act" (Presentation, Standing Committee on Aboriginal Affairs and Northern Development, Ottawa, ON, 5 December 2016), 8, https://sencanada.ca /content/sen/committee/421/APPA/Briefs/PamelaPalmater_2016-12 -05_e.pdf.

25 Palmater, "Bill C-3 Gender Equity in Indian Registration Act," 8–10.

26 Pamela Palmater, "Forcing our Hearts," *Fuse Magazine*, May 2012, 4.

27 Arthur J. Ray, Jim Miller, and Frank J. Tough, *Bounty and Benevolence: A History of Saskatchewan Treaties* (Montreal: McGill-Queen's University Press, 2000), 202.

28 Morris, *The Treaties of Canada*, 92–125.

29 Morris, 92–3.

30 Morris, 228.

31 Morris, 42.

32 Morris, 42.

33 Chris Andersen, *Métis: Race, Recognition, and the Struggle for Indigenous Peoplehood* (Vancouver: UBC Press, 2014), 40–1.

34 Andersen, *Métis*.

35 Kent McNeil, "Sovereignty on the Northern Plains: Indian, European, American and Canadian Claims," *Journal of the West* 39, no. 3 (2000): 12.

36 Morris, *The Treaties of Canada*, 98.

37 Robert Innes, *Elder Brother and the Law of the People: Contemporary Kinship and Cowessess First Nation* (Winnipeg: University of Manitoba Press, 2013), 72.

38 Innes, *Elder Brother and the Law of the People*, 71.

39 Innes, 72, 82.

40 Innes, 75.

41 Innes, 75–6.

Chapter Four

Sharing of the Dish: The Dish with One Spoon and Environmental Planning in Toronto

DA CHEN

Introduction

Toronto is a city of 2.93 million people located on the northern shore of Lake Ontario and the traditional territory of many nations, including the Mississaugas of the Credit, the Anishnabeg, the Chippewa[1], the Haudenosaunee, and the Wendat peoples. It's also covered by Treaty 13, Treaty 13a with the Mississaugas of the Credit, the Williams Treaty, and is also part of the Dish with One Spoon territories.[2]

In recent years, there has been growing anxiety regarding climate change and environmental issues in Toronto. With these concerns, the notions of sustainability and sustainable development have shaped planning discourse and have become essential goals of planning. However, in the environmental planning process in Toronto, there has been a lack of engagement with Indigenous people. Many existing Indigenous engagement processes are either tokenistic or ineffective, and the approaches utilized often do not allow Indigenous people to share their opinions.[3]

With the release of the final report of the Truth and Reconciliation Commission and its 94 Calls to Action, there is an urgent need for planners to at least understand and include Indigenous perspectives on environmental planning.[4] In June 2019, the Ontario Professional Planners Institute released a report titled *Indigenous Perspective in Planning*.[5] The report guided the institute to a better understanding of Indigenous peoples' perspectives on planning and how to strengthen the institutional framework so professional planning can more effectively support Indigenous planning approaches and perspectives. However, this should only be the first step, and planning as a profession must transform from its initial colonial project to one that reflects the values and needs of Indigenous people. As environmental planning always involves land, planners must have a better understanding of the various treaty obligations that are required and ensure the sharing and creating of mutual benefits.

Figure 4.1. This photo is from Da's 2017 trip to the Arctic, where he learned from Indigenous elders and youth about the importance of Indigenous knowledge and understanding the world from a different perspective.

Source: Courtesy of Da Chen.

Many Indigenous groups see self-determination as having the right to identify their problems and making choices to their goals in policy processes.[6] This approach challenges Canadian federalism and Canada's assertion that the federal and provincial governments have jurisdiction and responsibilities over Indigenous communities. This is part of the inherent tensions in the inclusion of treaties such as the Dish with One Spoon in the planning process as "any transformative process is by definition very political" and requires planning as a process to "discuss, debate, mediate and negotiate" with the different values and world views involved.[7]

In this essay, I will argue that one way planners in Toronto can better facilitate environmental planning is by engaging with the Dish with One Spoon.

The Dish with One Spoon is an important treaty that governs the relationships between Indigenous people and non-Indigenous people on this land. As stated by Mohawk scholar Ruth Koleszar-Green, many of us are guests on this land and have a responsibility to "learn about the history and current story of the land, to listen and learn protocol and to unsettle the privilege of ignorance."[8] To truly address the environmental crisis, environmental planners in Toronto need to understand the principles of the Dish with One Spoon and implement them in environmental planning processes. As planning has historically been complicit in state colonialism by providing the intellectual, conceptual, and technical skills to facilitate the clearance of Indigenous people, this means that planning has a responsibility not just to confront its complicity but to also aid in the recovery and re-inclusion of Indigenous communities.[9]

Approach

Situating the Work

As I embark on this research, it is important to acknowledge my positionality, my own biases and world views. I am a first-generation immigrant of Chinese descent. My Chinese Canadian identity shaped a lot of my upbringing, and it helped inform much of my understanding of the world and my relationships to it. Being raised in an immigrant household, I was taught to respect authority and to be grateful for the opportunity to live in a country like Canada. My early upbringing created an idealistic perception of Canada as an open, liberal, and inclusive nation.

Growing up in Toronto and being educated in Western institutions have also significantly moulded my understanding of Indigenous peoples, their cultures, and histories. Like many Canadians, I was taught to think about Indigenous peoples through the stereotypes often perpetrated by the Canadian government and mainstream media. As I started my post-secondary education at the University of Toronto Scarborough, and through my educational travel to the Arctic, I learned about Canada's history from the perspectives of Indigenous elders and youth. These learnings meaningfully reshaped my understanding of Canada and my role as a guest/settler. I learned about the broken treaties and the attempts by the Canadian government to implement policies to exploit, assimilate, and eradicate Indigenous people. Working on this research, my status as a non-Indigenous researcher was also something I had to come to terms with. I had to understand my role and positionality and ask myself a series of questions, like those mentioned in Leanne Simpson's book *As We Have Always Done*.[10] I had to ask myself whether I can use the knowledge that was shared with me in an ethical and appropriate way, given the colonial context within which scholarship is taking place. I had also to be mindful of whether my research could perpetuate colonialism. Additionally, I had to ensure that my research does not aim to speak on behalf of Indigenous people, but rather conveys the perspective of a guest/settler attempting to understand his role and obligation to treaties.

This research is motivated by my position as a future planning practitioner and my past experiences learning from Indigenous elders. I especially remember the words of an Indigenous speaker at a 2017 conference in Toronto. She mentioned that the reason that our cities are broken is that the planners themselves are broken inside. She insisted that before we can plan the ideal city, we must first transform ourselves. Her words that day reminded me of the importance of recalibrating and transforming our relationship to nature and each other. As many planners are educated in institutions based on capitalistic and white perspectives, I want to do my part and be responsible for educating others on reconciliation. As I am a guest on this land, I have a responsibility to learn about this place, its history, its current story, and to recognize my privilege and to use it in a way that gives back to the community.

Through this research, I also grapple with many tensions that exist, such as how to ensure Indigenous methodology and values are reflected in the planning process. I also had concerns with the existing planning profession's attempt to incorporate Indigenous values and world views into its pre-existing structure, as opposed to truly addressing the underlying issues. These tensions continue to emerge throughout my research and are something I am still attempting to understand and grapple with throughout my personal life and career.

I believe that it is vital for me to acknowledge these relationships and biases as I continue this journey of learning and understanding my obligation as a guest on this land.

Research Question and Framework

Many papers have been written on the need to include Indigenous people and Indigenous knowledge in the environmental planning process, but few focus on the role of treaty obligations in this process. The purpose of this essay is to explore and understand the roles of planners in fulfilling the obligations of the Dish with One Spoon. As Toronto is located on territories covered by this treaty, environmental planners should have a duty to plan according to the guidelines and principles laid out in it.

To this end, this essay will aim to address three related questions:

1. What is the Dish with One Spoon?
2. Why should the Dish with One Spoon inform environmental planning?
3. What are the environmental obligations represented in the Dish with One Spoon?

Literature Review

Most existing literature on the Dish with One Spoon focuses on issues of treaties, Indigenous planning, colonialism, dispossession, and Indigenous sovereignty. While most literature I reviewed doesn't look specifically at the Dish with One Spoon, it provides essential historical and contemporary insight into planning and the ongoing dispossession of Indigenous land in Canada. It has also allowed me to gain a better understanding of the barriers and challenges of existing planning practices in fulfilling the obligations under the Dish with One Spoon.

The following sections will address (in this order) the topics of treaty and treaty rights, existing planning practices, environmental planning, Indigenous planning and knowledge, and, lastly, the Dish with One Spoon. These sections are organized in this way first to establish the context in which the Dish with One Spoon is significant, such as treaties, colonialism, and dispossession. This is essential literature that readers need to know as, without this background, it would be difficult to grasp the significance of the Dish with One Spoon in the area of planning.

Treaty and Treaty Rights

According to the Idle No More movement, treaties are nation-to-nation agreements between the various Indigenous nations and the Crown.[11] In the understanding of many Indigenous peoples, treaties are agreements that cannot be altered or broken by any party. The spirit and intent of the treaty agreements meant Indigenous peoples would share the land but retain their inherent rights to lands and resources.[12] Indigenous signatories also understood treaties as a relationship with both rights and responsibilities,[13] and when they signed treaties, this was not understood as giving up their sovereignty, independence, or nationhood. Indigenous people have always argued that the treaties signed between the nations outlined the rights and responsibilities of both parties within this relationship and demonstrated that the Indigenous nations did not intend for their communities to be subsumed by the British Crown or subsequently the Canadian state.[14]

Under section 35 of the *Constitution Act, 1982*, "existing Aboriginal land rights can no longer be extinguished without the consent of those Aboriginal Peoples holding interests in those lands. Aboriginal consent may be required to give effect to legislation purporting to extinguish Aboriginal land rights, even if compensation is paid."[15] It's important to note this, since while the government of Canada has stated that treaties provide a framework for living together and sharing the land with Indigenous peoples, as well as the foundations for ongoing cooperation and partnership, it has also failed to acknowledge that Indigenous sovereignty and rights are the basis of treaty relationships.[16]

Additionally, treaties are grounded in the world views, knowledge systems, and political and cultural traditions of the nations that were involved and governed by Indigenous ethics of justice, peace, respect, reciprocity, and accountability.[17] Indigenous scholar Leanne Simpson and Canadian scholar Shiri Pasternak have discussed the historical processes Indigenous nations have for making treaties and maintaining a peaceful relationship with the land and each other. Simpson mentions Gdoo-naaganiaa, which means "Our Dish," and refers to the pre-colonial treaty between the Nishnaabeg and the Haudenosaunee Confederacy. This treaty is one of friendship and set forth the terms for taking care of the shared territory and was designed to promote peaceful coexistence between two sovereign nations.[18] Pasternak also refers to the concept of Onakinakewin, which was practised by the Algonquin of Barriere Lake.[19] Onakinakewin represents the Algonquins' understanding of the land, and it is a sacred constitution that contains the law that governs all living relationships in their world.[20]

However, these principles, guidelines, and laws have not been recognized by the colonial state, and, problematically, many treaties signed between Indigenous and European nations were often based on the premises of sovereignty that reflects only the European conception and the colonizer's ultimate dominion over the land based on the principle of *terra nullius*.[21] The adversarial nature of western European legal tradition demonstrates violence to the Aboriginal understanding of political sovereignty. The claiming of Canada is founded on the respect for the supremacy of God, and the rule of law is a non-cultural statement.[22] Even when treaties such as the Oregon Treaty were created between European nations regarding Indigenous land, Indigenous nations were often not included. In the eyes of the colonial powers, the issue of sovereignty was settled, and they had decided that Indigenous peoples did not possess sovereignty over these lands.[23] However, in Canada under the Royal Proclamation of 1763, *Aboriginal title* was explicitly understood to ensure that all land would be considered Aboriginal land until ceded by treaty.[24] As many territories in Canada are unceded, this raises questions about the sovereignty of Indigenous peoples. The Royal Proclamation should be used to buttress assertions of Indigenous sovereignty, yet in the contemporary planning process, it is rarely mentioned.

When discussing treaties, it is also essential to understand treaty responsibilities. Many scholars argue that treaty rights go beyond the relationship between the Crown and Indigenous governments and is something every member of a signatory nation has a duty and obligation to fulfil.[25] Recognizing our treaty responsibilities is an essential step in addressing Toronto's own colonial legacies and moving forward to rebuild the relationship between settlers and Indigenous peoples in the area. According to scholar Kirke Kickingbird, "treaties form the backdrop of the past, confirm rights in the present and provide the basic definition for the evolving future."[26] Pasternak also highlights that it's essential to know the source of the jurisdiction of the colonial state.[27] For the colonial state of Canada, jurisdiction focuses on the accumulation and transaction of land and resources. However, this contrasts with Indigenous legal orders, which focus on ensuring the care of the land and people, and which are reinforced in their understanding of the treaties.[28] The Dish with One Spoon focuses on ensuring there are enough resources for future generations; environmental planners should thus ensure that the planning process is reflective of Indigenous legal orders and respect the treaties and Indigenous sovereignty.

Existing Planning Practices

In Ontario, planning falls under the jurisdiction of the provincial government, which frequently delegates the responsibility of land-use planning to municipalities such as the City of Toronto. The Canadian Institute of Planners defines planning as the "scientific, aesthetic, and orderly disposition of land, resources, facilities and services to secure the physical, economic and social efficiency, health and well-being of urban and rural communities."[29] It has been defined in similar ways by other professional institutes of planners internationally. Planning as a profession has been complicit in the erasure of Indigenous knowledge and sovereignty and has struggled to accommodate and comprehend the challenges faced by Indigenous peoples.[30] This is because planning has been actively involved in the dispossession, oppression, and marginalization of Indigenous people to make way for colonial resettlement and economic projects.[31] The aim of these colonial projects has always been to clear the way for the settler state, and planning is complicit in this by providing the intellectual, conceptual, and technical skills to facilitate the clearance of Indigenous people.[32] Going forward, planners have a responsibility not just to confront their complicity but also to aid in the recovery and re-inclusion of Indigenous communities.[33]

In recent years, with a series of court rulings such as the *Haida Nation v British Columbia (Minister of Forest)*, the duty to consult and accommodate has become one of the essential principles of Canada Aboriginal law.[34] However, even with these principles, most engagement processes are still ineffective when consulting Indigenous people. While the duty to consult is grounded in the honour of the Crown, which is a core principle that informs all interaction between Indigenous people and the government, the same legal obligation does not apply to municipalities.[35] The limitation of the duty to consult in municipal planning decisions can be seen in the *Neskonlith Indian Band v Salmon Arm (City of Neskonlith)*.[36] In that court case, the British Columbia Court of Appeal ruled that as municipalities are not entities of the Crown, they do not have an obligation to consult.[37]

This shows that the principle of duty to consult is insufficient when working with Indigenous peoples, and that planners need to transform planning practices so as to better reflect the world views and needs of Indigenous communities.

Environmental Planning

Currently, it can be argued that the environmental planning process in Toronto is based on the idea of sustainable development. With growing concern regarding climate change and environmental degradation, the concepts of sustainability and sustainable development have appeared in various reports, such as *Toronto's First Resilience Strategy* and *TransformTO*, and has become an essential part of planning.[38] Sustainability is generally understood as "a guiding framework for attempts to change politics, economy, and society towards enabling more sustainable ways of living."[39] The dominant sustainability discourse tends to reflect the recommendations of the 1987 Brundtland Report, which viewed "economic development as essential to meeting social goals of sustainable development."[40] While many planning scholars highlight the importance of sustainability and its inclusion in planning, others are more critical. Some have argued that the current concept and understanding of sustainability is flawed and has a negative influence on planning.[41] As stated by Indigenous scholar Robin Kimmerer, sustainable development sounds like continuing as we always have and maintains a focus on "What we can take?" rather than "What can we give to Mother Nature."[42] With the existing dominant Western world view, society is still predicated on the hope of a better future through a materialistic understanding of growth and economy.[43]

For many scholars, to address the existing gap in the environmental planning process, there needs to be a focus on transforming the current planning practices to reflect Indigenous world views and values. Principles such as reciprocity, relationships, and responsibilities are thus highlighted. Anishinaabe scholar Deborah McGregor emphasizes the importance of reciprocity, relationship, and responsibility.[44] She indicates that reciprocity is about how knowledge is shared, the conditions in which it is shared, and why it is shared. Relationship and responsibility aim to ensure respectful and mutually beneficial relationships are not neglected.[45] These are all values that are important to creating an equal relationship between Indigenous and non-Indigenous people.

Historically, there has been a lack of engagement with Indigenous world views in environmental planning processes, such that Western knowledge is often allowed to overshadow local Indigenous views and to deny the validly of Indigenous knowledge.[46] The European conception of orderly development has rejected Indigenous agriculture and environmental management as irrational and unproductive, and the legal system continues today to be structured around white privilege and priorities.[47] In many cities, this occurs through the "construction and separation of Indigenous and non-Indigenous spaces by the state, and this process continues to happen as the state continues to police Indigenous place-making and self-determination, particularly in relation to cities."[48] This process often occurs through restricting Indigenous people's land use, active erasure of Indigenous history, and a series of institutional mechanisms that erase Indigenous titles and rights in cities.[49] There also exist deep-seated tensions regarding historical rights and access to land and resources, and there are many times when Indigenous jurisdiction or constitutionally enshrined titles and rights go unrecognized during consultation and engagement processes.[50]

Indigenous Planning and Knowledge

According to scholar Ted Jojola, Indigenous planning is defined as both an approach to community planning and an ideological movement.[51] He states that what distinguishes Indigenous planning from mainstream practice is its reformulation of planning approaches in a manner that incorporates "traditional" knowledge and cultural identity. A key component of this process is the acknowledgement of Indigenous world views and how these distinguish Indigenous communities from neighbouring non-land-based ones. Many scholars see the ultimate aim of Indigenous planning to be improving the lives and conditions of Indigenous people and to "refuse" ongoing exploitation, oppression, and extinction.[52] Indigenous knowledge is often built over millennia and provides the basis for the understanding of resources, their availability and temporal variability.[53] The failure to engage and recognize Indigenous knowledge often limits the capacity of Indigenous people to be involved in the environmental planning processes.

Indigenous knowledge is essential and can help address pressing environmental issues as they are rooted in particular areas and are practical and collective. Indigenous planning is often "community/kinship- and place-based and is rooted in specific Indigenous people's experiences linked to specific places, land and resources,"[54] and through working and empowering Indigenous communities, we can create the most sustainable and grounded planning processes.[55] Some scholars, such as John Borrows, even argue that "North American democratic institutions should more effectively link democracy and environment and provide for the participation of Indigenous people."[56] He believes that by better engaging Indigenous people, we can design a more sustainable future, and refers to the Hay Island proposal as an example of how this could occur. As Indigenous knowledge is predicated on a better understanding of and respect for others, to be effective, planners need to be culturally appropriate and respectful of Indigenous cultures rather than imposing Western ideas and practices.[57] Most importantly, Indigenous planning recognizes that Indigenous people are not stakeholders but rights holders and that they have the right to self-determination as stated under the UN Declaration on the Rights of Indigenous Peoples.[58]

The Dish with One Spoon

According to most historical records, the Dish with One Spoon is usually understood as a treaty that was renewed throughout the seventeenth, eighteenth, and nineteenth centuries between Algonquian and Iroquoian nations and, subsequently, Europeans and all newcomers.[59] In most interpretations of the treaty by Indigenous scholars, the dish symbolizes the common hunting ground shared among the inhabitants. At the same time, the spoon denotes that all people who share the territory are expected to limit the resources they take and leave enough for others. The Dish with One Spoon specifies that each nation is to share resources and land responsibly, taking only what is needed and not more than what can be sustained by nature. According to the Haudenosaunee records, the Dish with One Spoon is one of the most critical responsibilities outlined under the Great Law. The principle of the Dish with One Spoon states that

> It will turn out well for us to do this: we will say, "we promise to have only one dish among us; in it will be beaver tails, and no knife will be there."

Thereupon the chiefs confirmed that so it should happen. Thereupon [the Peacemaker] said, "now we have completed the matter, we will have one dish, which means that we will all have equal shares of the game roaming about in the hunting grounds *hutowaaestahkwahek* and fields *kahetayetu'* and then everything will become peaceful among all of the people, and there will be no knife near our dish, which means that if a knife were there, someone might presently get cut, causing bloodshed, and this is troublesome, should it happen thus, and for this reason, there should be no knife near our dish."[60]

As highlighted throughout this literature review, it's important to recognize role of treaty rights and the history of dispossession and erasure of Indigenous people. While the Dish with One Spoon contains many principles that could help transform environmental planning, it's essential for planners to realize that it's also a treaty that encompasses Indigenous laws and world views.

Research Methods

Key Informant Interviews

For this research, semi-structured interviews were conducted with six key informants alongside an analysis of the literature. The primary research method was through interviews with planners, policymakers, and Indigenous knowledge holders who have experience working on projects regarding treaty obligations. The primary information I sought and collected related to whether or not, how, and to what extent planners currently engage with the Dish of One Spoon treaty in the environmental planning process. Additionally, this research focused on understanding if or how the Dish with One Spoon treaty affects planning policies. At the same time, the literature review provided a picture of the theoretical framework around treaties, dispossession, and colonialism. Each interview was approximately one hour in length, and the people I interviewed represent different organizations and viewpoints and include Indigenous scholars and knowledge holders, City of Toronto staff, representatives of NGOs, and a Toronto city councillor. They are:

- City Councillor Mike Layton
- Clara MacCallum Fraser, Shared Path Consultation Initiative
- Elder Philip Cote, First Story Toronto
- Dr. Alan Corbiere, York University

- Dr. Jon Johnson, University of Toronto
- Indigenous Affairs Office staff

The key informants were selected for different reasons. I wanted the interviews to reflect the diversity of those involved in this field and, at the same time, different perspectives and viewpoints. The main themes covered in the interviews range from the principles of the Dish with One Spoon, Indigenous world views, treaties, wampum belts, Indigenous history, the roles and responsibilities of those living on the land, education, and planning tools. The interviews not only helped fill in the gaps that existed in the literature and provided a more in-depth understanding of the Dish with One Spoon, but they also raised new questions and revealed some of the tensions in Indigenous and non-Indigenous people's understanding of the Dish with One Spoon.

Limitations

Certain faults may arise from using Western methodology to analyse the Dish with One Spoon. Indigenous methodologies are shaped by Indigenous paradigms, world views, and principles.[61] This means that they are influenced by Indigenous culture, socialization, and experiences. By utilizing Western methodology in the form of a literature review and interviews, I face challenges in sharing the Indigenous knowledge and stories that were shared with me. As I am a non-Indigenous person, I am not in a position to speak about some of the issues that emerged during the interviews. Additionally, the underlying matter of how planning could create the necessary space for Indigenous world view and values to transform the profession remains unresolved in this research. There are also tensions arising from planning's attempt to incorporate Indigenous values and world views into the current planning paradigm without acknowledging the questions surrounding land, dispossession, sovereignty, and treaties. Those are important issues that need to be better addressed in future research.

Key Informant Responses

Key informants provided valuable context and information on the Dish with One Spoon and how it may impact environmental planning. While some variations existed in the interviews, mostly concerning the relevance of the treaty, the Dish with One Spoon is generally described as a treaty or covenant initially between the different Haudenosaunee nations and the Anishinaabe, Mississaugas, and other nations aimed at ensuring that the environment is protected.

Elder Philip Cote, a traditional wisdom keeper and historian from Moose Deer Point First Nation, stated that the Dish with One Spoon has existed for thousands of years and governs relationships on this land. He describes it as a covenant that represents all the waterways and land around southern Ontario.[62] Similarly, according to Dr. Alan Corbiere, an Anishinaabe historian, the Dish with One Spoon is a set of overarching principles, and many similar treaties exist in other parts of North America and govern the relationship of many nations throughout the continent.[63] Additionally, Dr. Corbiere stated that there is a philosophical aspect of the Dish with One Spoon as it is based on the idea that one is not to be greedy and take more than one's share but must instead leave enough for future generations. As the land is understood as a dish, it is essential to ensure that a spoonful is all one takes from the dish at a time to ensure its sustainability for the future.[64] The insights from these interviews provided important context for how the Dish with One Spoon could impact environmental planning processes. They also highlighted that the Dish with One Spoon speaks of a political relationship between Indigenous and non-Indigenous people. Throughout the interviews, several themes and challenges emerged when discussing the role of planners and decision makers in considering the principles of the Dish with One Spoon and eventually transforming the environmental planning processes in light of the treaty.

Themes

THE RELEVANCE OF THE DISH WITH ONE SPOON

The relevance of the Dish with One Spoon in the environmental planning process in Toronto was discussed in many of the interviews and from a variety of perspectives. According to a Toronto Indigenous Affairs Office staff member, the Dish with One Spoon is only a historical treaty that was signed hundreds of years ago and does not currently have any legal impact on modern environmental planning. That staff member further elaborated that while many nations have historical claims to Toronto, such as the Huron-Wendat and Six Nations of the Grand River, the only treaty that is officially recognized by the Government of Canada is the Treaty 13 (1805) with Mississauga of the Credit First Nation.[65] This interpretation of the treaty means that while the Dish with One Spoon may have important principles and guidelines embedded within it, the City of Toronto has no legal obligation to implement them.

In addition to not having legal status, the Dish with One Spoon has a limited relevance in the environmental planning process, impacted as it is by the limitation of existing planning practices. City Councillor Mike Layton stated that the Planning Department at the City of

Toronto mostly deals with policy and has less flexibility on issues of land-use planning. This means that the city is limited in addressing issues around treaties and sovereignty. Additionally, Councillor Layton stated that many of the decisions concerning the environment in Toronto also happen outside of the city's jurisdiction and that, "unless we have other levels of government that are taking it [the environment] seriously and changing how decisions are made to ensure the voices of the environment has some power, then not much is going to change."[66] This means that even if the city wishes to implement the principles contained within the Dish with One Spoon, it does not have the political authority to do so.

Regarding the specific details of the Dish with One Spoon, the key informants generally agreed that the treaty provides an overarching guideline that reflects an Indigenous world view. However, as the treaty also lacks any specific guidelines that people can follow, it does raise concerns about how planners can fulfil the treaty's obligations. As there are no measurable criteria governing whether planners have fulfilled the treaty or not, this makes it difficult to measure whether environmental planners in Toronto are in accord with the treaty's obligations during the environmental planning processes. However, planners could look to the Royal Proclamation, Treaty of Niagara, section 35 of the *Constitution Act*, or section 25 of the *Charter of Rights and Freedoms* for guidance. While these treaties and laws are not perfect, they contain principles and mechanisms as defined by the courts that are legally binding and can be implemented to some extent.

However, other key informants argue that the Dish with One Spoon is still relevant and should continue to guide and inform modern planning practices. Dr. Corbiere stated that while the Dish with One Spoon may lack many of the clauses of a modern treaty, its relevance should not be dismissed:

> The Dish with One Spoon is a treaty. It may not have specific clauses the way a Western treaty does, which talks about specific clauses and demarcations of which territory or boundaries are set, nor does it specify what nations can only take X amount of deer. It doesn't have that, but it has the flexibility it's supposed to have. It has many principles that would be called co-management in the contemporary sense where the nations are supposed to work together to maintain the dish to ensure that there's something for future generations.[67]

As stated by Dr. Corbiere, the relevance of the Dish with One Spoon should not be dismissed as it contains many principles applicable to contemporary environmental planning practices, such as co-management, and requires the different nations to work together. Dr. Jon Johnson also stated that while the Dish with One Spoon is a historical (pre-Confederation) treaty, it's still relevant and has real implications on current practice:

> If they [early settlers] didn't agree to the Dish with One Spoon, then they wouldn't be allowed or invited to stay on the lands. Additionally, in the early period, settlers engaged in sustainable relationships with Indigenous peoples around the Great Lakes, and this early relationship should be evidence that indicated that the early settlers knew about the Dish with One Spoon and implicitly or explicitly agreed to the terms. As we are the descendants of those Indigenous and non-Indigenous people who made the agreement, or we have immigrated to the territory, I believe that the Dish with One Spoon is operative to all of us who currently live and work in this region.[68]

The two contrasting perspectives on the relevance of this treaty suggest that many gaps and tensions exist in terms of how the Dish with One Spoon is currently understood. This difference in views is significant as it speaks to broader structural tensions and issues, even if these are outside the scope of this research.

Indigenous World Views

The inclusion of Indigenous knowledge and world views in planning practices was seen as a major challenge by Clara MacCallum Fraser, who works with both planners and Indigenous communities in her work with the Shared Path Consultation Initiative. Fraser mentioned that one of the most challenging things is that many treaties, including the Dish with One Spoon, have a spiritual component that many planners and policymakers are very uncomfortable with. She recognized the hypocrisy of many planners' belief in secularism and highlighted the religious foundation of contemporary planning practice:

> We live in a society that often thinks of ourselves as secular and having separated our policies from spiritual or religious practices. However, spiritual or religious foundations play a vital role in shaping our ideas around land use, particularly for the early settlers arriving in North America. Those settlers had this view of the landscape and thought of the place like Eden, untouched and perfect for their use. It's important to acknowledge that these religious ideas always played a role in the formation of Canada, and with increasing environmental disaster throughout the world, it's essential to look back and re-evaluate our relationship to nature, a part of which is spiritual.[69]

Fraser also mentioned that the way we understand treaties such as the Dish with One Spoon differs from Indigenous peoples' understanding. Our usage of words such as "resources" reflects a fundamentally different understanding of nature and our way of thinking and knowing reflects a different understanding of the world. As Indigenous world views tend to focus on a more holistic understanding of the world that emerged from thousands of years of existence and experiences, while a Western world view tends to be more compartmentalized, this can be seen in our understanding and interpretation of treaties such as the Dish with One Spoon:

> When these treaties were signed between the Indigenous people and the settlers, while the words on the treaties might be agreed on by both parties, they may also have different interpretations of the treaties. Sometimes it's overwhelming for us to try to understand what treaties mean for current-day planners, and it's something I'm trying to achieve at work by looking at how I can help planners understand and translate treaty language.[70]

However, it's also important to be careful when attempting to contextualize Indigenous knowledge. As explained by scholar David Delgado Shorter, Indigenous knowledge must be understood in the context of Indigenous legal and political orders.[71] By framing Indigenous knowledge as spiritual, we might inadvertently depoliticize Indigenous knowledge and undermine its importance in the discourse around treaties and sovereignty.

Elder Philip Cote also mentioned that in Indigenous world views, there are many stories that focus on not being greedy and only taking what is necessary. These lessons provide important guidance about how Indigenous people should live and interact with nature. Additionally, he mentioned that there is an intricate cosmology embedded in the treaties and that this represents both the spiritual and physical presence of Indigenous peoples on the land.[72] It is important for planners to recognize *ontological pluralism*, the idea that there are different ways or modes of being, as there is a risk that the failure to recognize different world views can result in planners and policymakers omitting Indigenous concerns, such as Indigenous rights and sovereignty, during environmental planning processes.[73] As a researcher, I find it difficult, even uncomfortable to bring Indigenous world views in line with existing planning parameters. This is an ongoing tension that exists within planning. As Indigenous values and the current planning practices are fundamentally different, without radically transforming planning, any inclusion or incorporation of Indigenous ideas would only serve the colonial project.

Changing Existing Planning Processes

While there are many inadequacies and limitations within the existing environmental planning process in terms of engaging with Indigenous communities, there are also many steps in place to address some of these shortcomings. According to the staff member from the Toronto Indigenous Affairs Office, a multitude of initiatives are in the works that could improve existing environmental planning processes.[74] One involves setting up a new planning council that would include representation from urban Indigenous communities, architects, planners, and elders. The council would be there for urban planners and other relevant city development divisions to reach out to for guidance when planning for Indigenous people. Councillor Layton also believes that the Parks and Recreation Department at the City of Toronto could aspire to incorporate elements of the Dish with One Spoon. He stated that, ideally, the principles of the Dish with One Spoon could help guide the planning of parks in Toronto.[75] As parks spaces service the needs of various communities, the Dish with One Spoon can help planners gain a better understanding of the natural system and plan parks for present and future generations.

On the surface, it might seem relatively straightforward to include the principles embedded in the Dish with One Spoon in the environmental planning process. As stated by Dr. Johnson, the core values of the Dish with One Spoon are peace, equity, sustainability, and peaceful relationality with others. However, this could be much more challenging and difficult in practice. As there is a significant difference between Western and Indigenous world views, planners' understanding of the Dish with One Spoon might significantly differ from the treaty's original intent. Additionally, while the principles of the Dish with One Spoon may seem very similar to many existing environmental planning values, there are some fundamental differences. The major difference is that the Dish with One Spoon reflects Indigenous world views and encompasses all of creation, as opposed to just humans; consequently, if planners abide by the principles of the Dish with One Spoon, then they must think of all of creation. Without addressing this underlying tension, it would be challenging for any policies to truly reflect the principles and values of the Dish with One Spoon.

One big challenge remains. As stated by Clara MacCallum Fraser, how can planners become more engaged with treaties? She struggles to translate Indigenous world views and knowledge into planning vernacular. She questions planners "who are currently working in the city ... [but who haven't] even pondered their part in this story of truth and reconciliation ... who are just reading the Provincial Policy Statement."[76] She wonders how such planner can be more aware of treaties and Aboriginal treaty rights. While some progress has been made in including more Indigenous voices in environmental planning processes, many gaps still exist. The planning process doesn't currently recognize or reflect Indigenous world views, and there isn't any mechanism in place to truly reflect the principles of the Dish with One Spoon. Additionally, there is a critical need for planning to think differently. As stated by Fraser,

> If we are thinking about seven generations instead of the next twenty years, that could change the way planning could be done. However, we are working within systems that generally don't think that way. For us [planners] to actually plan for seven generations requires radical change, which is something that is not happening yet.[77]

Fraser is here invoking the Seventh Generation Principle, which is based on an ancient Iroquois philosophy that "the decisions we make today should result in a sustainable world seven generations into the future."[78] This means that planners need to rethink planning in a way that reflects Indigenous values and world views. It also means that any decisions related to land or the environment need to result in a sustainable environment for those living seven generations into the future.

Education and Decolonizing Planning

To truly transform planning and to have an environmental planning process that reflects the values of the Dish with One Spoon, there is a vital need to educate current and future planners. As stated by the staff member at the Indigenous Affairs Office, there should be more opportunities for the Indigenous communities in Toronto to connect with universities and set up institutes of planning. Planning organizations should also create educational opportunities for planners to become more aware of Indigenous issues and to act to create the necessary changes. He stated that "if planners are educated on treaties throughout planning school and [in their] professional careers, then it would go a long way in addressing some of the gaps that currently exist." He recognizes that, historically, planning as a profession has not done a great job recognizing treaties or working alongside Indigenous people. Still, he has seen positive changes in the profession in recent years, such as the Ontario Professional Planners Institute's creation of an Indigenous Planning Perspective Taskforce in 2018 and their subsequent publication of a report on the role of Indigenous perspectives in planning.[79] Additionally, he mentioned the importance of developing toolkits for planners. As planning follows a very specific paradigm, and is highly complex, there needs to be toolkits that allow planners to understand the impacts of historical processes on the present-day Toronto. He said,

> I'm also saying we need to give people practical tools. So how do you engage a community? How to better recognize treaties? Who should planners be speaking to? And many other kinds of practical issues. What I'm proposing in the system is that you take knowledge, like applied knowledge, and create a tool where [that] knowledge [is] passed to other planners. It becomes something that planners are going to use pretty much every day.[80]

However, while a toolkit might be important in addressing some of the gaps that currently exist and providing planners with the tools to better engage and work alongside Indigenous people, these tools are still limited in many ways. A toolkit would be insufficient in addressing the need for ongoing conversations surrounding the renewal of nation-to-nation relationships or questions of self-determination as it would not have the flexibility necessary to answer these important questions on responsibility and reciprocal relationships.

Additionally, it is also crucial for planners to engage and learn from Indigenous leaders. These leaders would provide valuable knowledge and guidance to planners during the environmental planning processes. The same member of the Indigenous Affairs Office said,

> The planner should be working with the Indigenous leaders of the area, and by leaders, I don't just really mean the elected leadership, but the elder, like we used to have. Elders Councils, Youth Councils, Women's Councils. So, planners first should ask those people, the elders and the ceremony people, and tell them what they want … and whether it's the right thing to do. And then, the [Indigenous] people deliberate upon it, and they would then inform the planners that this is what they should do and here are some of the constrictions.[81]

Learning from Indigenous elders and educating current and future planners would go a long way towards addressing the existing gaps within environmental planning. It can be argued that there are currently a lack of Indigenous voices and world views within the planning process; moreover, many processes that do involve Indigenous people are limited and often do not reflect the needs or values of Indigenous nations. With more knowledge and awareness of the Dish with One Spoon and other relevant treaties, planners can access more tools and have a greater capacity for addressing some of the underlying tensions in the environmental planning process. The interviews reflect Matunga's observation that "there is an important need for planners to accept the legitimacy of Indigenous planning as a parallel tradition that has its own history, focus, goals and approach."[82] This means that planners shouldn't merely incorporate Indigenous knowledge or ideas into existing practices; rather, they should seek to reorganize and indeed transform planning as a profession in such a way that would allow these different ideas to exist.

Challenges

Researching as a Non-Indigenous Person

While embarking on this research as a non-Indigenous person, I came across many tensions and issues around Indigenous knowledge and treaties that I did not feel comfortable addressing. Regarding treaties, while the Crown only recognizes Treaty 13 with the Mississauga of the Credit First Nation, there are existing tensions with other nations who have historical claims, such as the Six Nations of the Grand River and the Huron-Wendat Nation. Additionally, as Toronto is home to a large urban Indigenous population, questions remain as to whether the city is legally obliged to consult them on matters of treaties and sovereignty.

Elder Philip Cote stated during one of the interviews that Indigenous culture is something that is experienced; this raises questions about how I can translate such oral knowledge into an academic essay. I felt discomfort in trying to articulate something that cannot be understood in the current planning framework. While certain conceptual tools have emerged in past projects such as the Hay Island proposal,[83] I am still learning of ways to translate this oral knowledge into a written format without undermining the values and essence of this knowledge. This is still a major challenge I am trying to confront as I continue this academic journey. This research raised many more questions for me regarding my role as a treaty person and a guest on this land.

Through this research, I was also able to gain a better understanding of my role as a treaty person. Through the literature reviews and interviews, I learned about ways to uphold my responsibility as a guest on this land and ways to build towards an inclusive and equitable future.

Planning as a Colonial Project

Throughout this research, I came across the challenges of bridging Indigenous and non-Indigenous world views. It can be argued that current planning practices are based on the world view of Euro-modernity, which can be described as capitalist, rationalist, liberal, secular, patriarchal, and white.[84] In this world view, planning is mostly interested in managing private property, and our relationship with nature and land is seen as extractive. This contrasts with Indigenous world views, which are more holistic and look at the relationship to both the human and non-human world. I found it difficult to include the principles of the Dish with One Spoon into existing planning practices. I see these practices as siloed, professional, and governed by Canadian laws and different jurisdictions. This view is challenged by the Dish with One Spoon, which speaks of an inherently different type of planning. The Dish with One Spoon reflects Indigenous world views and laws that are more holistic and reflective of all of creation, as opposed to just human. Instead of merely incorporating Indigenous values and world views as reflected in the Dish with One Spoon, it would be more important to transform planning as a profession. Merely embedding or including some principles of the Dish with One Spoon into existing planning practices risks romanticizing and depoliticizing Indigenous knowledge. This would only result in changes on the surface, leaving institutional tensions and barriers to remain. I implore current and future planners to recognize this and to work towards fundamentally changing planning.

Sets of Guiding Principles for Planning

The following are sets of guiding principles that emerged during the key informant interviews. These are important to highlight for future planning practitioners.

- The core values of the Dish with One Spoon are peace, equity, sustainability, and peaceful relationality with others and all of creation, as opposed to just humans. These values mean that in all environmental planning processes, planners need to assess whether planning processes fulfil the values of peace, equity, sustainability, and peaceful relationality, not just in the context of humans but with all creation. This guiding principle also emerged from the interviews; it is also drawn from the work of Anishinaabe scholar Deborah McGregor, who emphasizes the importance of practising reciprocal relationships with the Indigenous nations and focusing on building and maintaining a respectful relationship with these communities.[85] As stated by Dr. Alan Corbiere during the interview, the Dish with One Spoon "has the flexibility that would allow different nations to work together in taking care of the land."[86] This means that different groups should be able to understand the principles of the Dish with One Spoon and to work together to create a more sustainable and inclusive future.

- Planners should practise "seventh generation thinking," which means thinking about those living seven generations in the future. This is reflective of the Seventh Generation Principle, which dates to the Great Law of the Iroquois Confederacy. Currently, many planning decisions follow a timeline of twenty to thirty years. It would be necessary for future planning decisions to assess whether those living seven generations into the future would approve of the decisions that are being made now. As the Dish with One Spoon is premised on the acknowledgment that what we extract from the environment can cause harm and impede its ability to sustain people and future generations, it is essential to ensure that our shared world is left in a better place for future generations.

- We should know the history of the land before Western settlement and learn from Indigenous elders. As Indigenous people have lived here for thousands of years, planners need to maintain open communication and engage Indigenous communities during every step of the environmental planning process. During the interview with Elder Cote, he mentioned that knowing the history of the land allows us to connect to the land and to remember all of its history and communities.[87] Knowledge of the land is critical in maintaining a different world and a different way of knowing. A better understanding of history before European settlement also challenges the notion that Western knowledge is superior and allows planners to gain a better understanding of Indigenous world views and values. This requires planners to be opened-minded when learning from Indigenous elders and to ensure their practices reflect these teachings.

- It is important for planning schools and planning institutions to provide workshops and educational opportunities for planners to learn about treaties such as the Dish with One Spoon. The education process should involve more than just incorporating Indigenous world views into the existing planning institutions; indeed, it should seek to fundamentally transform planning. This means recognizing the importance of Indigenous law, Indigenous world views, and Indigenous methodologies and introducing these to planning students and practitioners. These educational opportunities can go a long way towards improving planners' capacity to engage with Indigenous people and can aid in decolonizing planning.

Conclusion

Ultimately, environmental planners in Toronto should plan according to the principles of the Dish with One Spoon. The Dish with One Spoon is a treaty that contains a set of principles that reflect Indigenous laws and world views. There is an essential need for planners to engage with the treaty, as we all have a legal obligation to uphold the principles contained within it. To achieve these necessary changes, planners must think about planning in terms of relationships, and about the needs of all of creation, human as well as non-human. Planners should also think about their roles as treaty people and reflect on how their work can transform planning as a profession.

While planners do have a challenging role in working within the limits of the existing planning framework, they nevertheless have an obligation to work towards transforming current planning by learning from Indigenous people and to truly understand their roles as guests on this land. The role of planners on the Dish with One Spoon territory is to mediate this set of political relationships and to work to address some of the existing tensions identified in this essay. Taking the Dish with One Spoon into account in the environmental planning process is only a starting point. Planners have a responsibility to ensure that all future environmental planning processes are based on this historical relationship of equity, peace, sustainability, and reciprocity with the land and all creation. However, planners need to also realize that Indigenous knowledge and world views are political and should not undermine their importance in the discourse of treaties and sovereignty.

Planners must think further about the factors that allow this relationship to be maintained in the right way, and how planning can provide tools to address the tensions that exist in settler/guest-Indigenous relationships and the conditions and terms of reference governing these relationships.

It's important to remember that this research is just a small first step, and that much more work and further research is required if we are to understand how planners can not only engage with the Dish with One Spoon but also contribute to the larger discourse on Indigenous world views and Indigenous sovereignty in the planning process.

NOTES

1 Mississaugas and Chippewa are both Anishinaabe peoples.
2 "Land Acknowledgement," City of Toronto, accessed 5 July 2019, https://www.toronto.ca/city-government/accessibility-human-rights /indigenous-affairs- office/land-acknowledgement/.

3 Hannah Escott, Sara Beavis, and Alison Reeves, "Incentives and Constraints to Indigenous Engagement in Water Management," *Land Use Policy* 49 (2015): 382–93.

4 Truth and Reconciliation Commission of Canada, *Final Report of the Truth and Reconciliation Commission of Canada* (Montreal: McGill-Queen's University Press, 2015).

5 Indigenous Planning Perspective Trask Force, *Indigenous Perspectives in Planning: Report of the Indigenous Planning Perspectives Task Force* (Toronto: Ontario Professional Planners Institute, 2019).

6 Joanne Heritz, "From Self-Determination to Service Delivery: Assessing Indigenous Inclusion in Municipal Governance in Canada," *Canadian Public Administration* 61, no. 4 (December 2018): 596–615. See also L. Joanne Heritz, "Municipal-Indigenous Relations in Saskatchewan: Getting Started in Regina, Saskatoon and Prince Albert," *Canadian Public Administration* 61, no. 4 (December 2018): 616.

7 Hirini Matunga, "Theorizing Indigenous Planning," in *Reclaiming Indigenous Planning*, ed. Ryan Walker, Ted Jojola, and David Natcher (Montreal: McGill-Queen's University Press, 2013), 24.

8 Ruth Koleszar-Green, "Cultural and Pedagogical Inquiry," *Cultural and Pedagogical Inquiry*, no. 2 (2018): 166–77.

9 Mantunga, "Theorizing Indigenous Planning," 3, 9; Libby Porter and Janice Barry, "Bounded Recognition: Urban Planning and the Textual Mediation of Indigenous Rights in Canada and Australia," *Critical Policy Studies* 9, no. 1 (2015): 22–40.

10 Leanne Betasamosake Simpson, *As We Have Always Done* (Minneapolis: University of Minnesota Press, 2017).

11 "An Indigenous-Led Social Movement," Idle No More, accessed 8 November 2023, https://idlenomore.ca/about-the-movement/.

12 "An Indigenous-Led Social Movement."

13 Leanne Simpson, "Looking after Gdoo-naaganinaa: Precolonial Nishnaabeg Diplomatic and Treaty Relationships," *Wicazo Sa Review* 23, no. 2 (2008): 29–42.

14 Simpson, "Looking after Gdoo-naaganinaa."

15 *Constitution Act, 1982*, Schedule B to the *Canada Act, 1982*, c 11 (UK).

16 "Treaties and Agreements," Indigenous and Northern Affairs Canada, last modified 11 April 2023, https://www.rcaanc-cirnac.gc.ca/eng /1100100028574/1529354437231.

17 Simpson, "Looking after Gdoo-naaganinaa."

18 Simpson.

19 Shiri Pasternak, *Grounded Authority: The Algonquins of Barriere Lake against the State* (Minneapolis: University of Minnesota Press, 2017), 77.

20 Pasternak, *Grounded Authority*, 78.

21 Patrick Wolfe, "Settler Colonialism and the Elimination of the Native," *Journal of Genocide Research* 8, no. 4 (2006): 387–409.

22 Laura DeVries, *Conflict in Caledonia: Aboriginal Land Rights and the Rule of Law* (Vancouver: UBC Press, 2011), 144.

23 Cole Harris, "How Did Colonialism Dispossess? Comments from an Edge of Empire," *Annals of the Association of American Geographers* 94, no. 1 (2004): 165–82.

24 "Treaties and Agreements," Indigenous and Northern Affairs Canada.

25 Chris Hiller, "No, Do You Know What Your Treaty Rights Are? Treaty Consciousness in a Decolonizing Frame," *Review of Education, Pedagogy, and Cultural Studies* 38, no. 4 (2016): 381–408.

26 Kirke Kickingbird, "What's Past Is Prologue: The Status and Contemporary Relevance of American Indian Treaties," *St. Thomas Law Review*, no. 7 (1995): 603–29.

27 Pasternak, *Grounded Authority*, 78.

28 Pasternak, 78.

29 "About Us," Canada Institute of Planning, accessed 19 March 2020, https://www.cip-icu.ca/About/About-Us.

30 Porter and Barry, "Bounded Recognition."

31 Magdalena Ugarte, "Ethics, Discourse, or Rights? A Discussion about a Decolonizing Project in Planning," *Journal of Planning Literature* 29, no. 4 (2014): 403–14; Harris, "How Did Colonialism Dispossess?"; Porter and Barry, "Bounded Recognition."

32 Mantunga, "Theorizing Indigenous Planning," 3.

33 Mantunga, "Theorizing Indigenous Planning," 9; Porter and Barry, "Bounded Recognition."

34 Haida Nation v British Columbia (Minister of Forests), 2004 SCC 73, [2004] 3 SCR 511 (Can.).

35 Shin Imai and Ashley Stacey, "Municipalities and the Duty to Consult Aboriginal Peoples: A Case Comment on Neskonlith Indian Band v. Salmon Arm (City)," *UBC Law Review* 47, no. 1 (2014): 293–331.

36 Neskonlith Indian Band v Salmon Arm (City), [2012], 327 BCAC 276 (Can. BC).

37 Imai and Stacey, "Municipalities and the Duty to Consult Aboriginal Peoples."

38 *Toronto's First Resilience Strategy*, City of Toronto, accessed 10 November 2023, https://www.toronto.ca/services-payments/water-environment/environmentally-friendly-city-initiatives/resilientto/; *TransformTO Net Zero Strategy*, City of Toronto, accessed 10 November 2023, https://www.toronto.ca/services-payments/water-environment/environmentally-friendly-city-initiatives/transformto/.

39 Daniela Gottschlich and Leonie Bellina, "Environmental Justice and Care: Critical Emancipatory Contributions to Sustainability Discourse," *Agriculture and Human Values* 34, no. 4 (2017): 941–53, https://doi.org/10.1007/s10460-016-9761-9.

40 Michael Gunder, "Sustainability: Planning's Saving Grace or Road to Perdition?," *Journal of Planning Education and Research* 26, no. 2 (2006): 208–21.

41 Gunder, "Sustainability."

42 Robin Wall Kimmerer, *Braiding Sweetgrass: Indigenous Wisdom, Scientific Knowledge and the Teachings of Plants* (Minneapolis: Milkweed Editions, 2013), 190.

43 Gunder, "Sustainability," 218.

44 Deborah McGregor, "Traditional Knowledge, Sustainable Forest Management, and Ethical Research Involving Aboriginal Peoples: An Aboriginal Scholar's Perspective on Traditional Knowledge, Sustainable Forest Management, and Ethical Research Involving Aboriginal Peoples," in *Aboriginal Policy Research: Voting, Governance, and Research Methodology*, vol. 10, ed. Jerry P. White, Julie Peters, Dan Beavon, and Peter Dinsdale (Toronto: Thompson Education Publishing, 2010), 227–44.

45 McGregor, "Traditional Knowledge," 237.

46 Brent Taylor and Rob C. De Loë, "Conceptualizations of Local Knowledge in Collaborative Environmental Governance," *Geoforum* 43, no. 6 (2012): 1207–17.

47 DeVries, *Conflict in Caledonia*, 67.

48 Julie Tomiak, "Contesting the Settler City: Indigenous Self-Determination, New Urban Reserves, and the Neoliberalization of Colonialism," *Antipode* 49, no. 4 (2017): 928.

49 Tomiak, "Contesting the Settler City."

50 Suzanne Von Der Porten and Robert C. De Loë, "Collaborative Approaches to Governance for Water and Indigenous Peoples: A Case Study from British Columbia, Canada," *Geoforum* 50 (2013): 149–60.

51 Ted Jojola, "Indigenous Planning – An Emerging Context," *Canadian Journal of Urban Research* 17, no. 1 (2008): 37–47.

52 Mantunga, "Theorizing Indigenous Planning," 5.

53 Escott, Beavis, and Reeves, "Incentives and Constraints."

54 Mantunga, "Theorizing Indigenous Planning," 5.

55 Taylor and De Loë, "Conceptualizations of Local Knowledge in Collaborative Environmental Governance."

56 John Borrows, "Living between Water and Rocks: First Nations, Environmental Planning and Democracy," *University of Toronto Law Journal* 47, no. 4 (1997): 417.

57 Escott, Beavis, and Reeves, "Incentives and Constraints."

58 United Nations, GA Res. 61/295, United Nations Declaration on the Rights of Indigenous Peoples (13 September 2007), https://undocs.org/A/RES/61/295; Von Der Porten and De Loë, "Collaborative Approaches to Governance for Water and Indigenous Peoples."

59 Victor P. Lytwyn, "The Dish with One Spoon: The Shared Hunting Grounds Agreement in the Great Lakes and St. Lawrence Valley Region," in *Papers*

of the Twenty-eighth Algonquian Conference, vol. 28, ed. David H Pentland (Winnipeg: University of Manitoba, 1997), 210–27.

60 Susan M. Hill, *The Clay We Are Made Of: Haudenosaunee Land Tenure on the Grand River* (Minneapolis: University of Minnesota Press, 2017), 42.

61 Kathleen E. Absolon, *Kaandossiwin: How We Come to Know* (Halifax: Fernwood Publishing, 2011).

62 Elder Philip Cote (traditional wisdom keeper, historian from Moose Deer Point First Nation), interview with the author, 17 January 2020.

63 Alan Corbiere (Anishinaabe historian), interview with author, 17 January 2020.

64 Corbiere, interview with author.

65 "The Toronto Purchase Treaty, No. 13 (1805)," Mississaugas of the Credit First Nation, 3 November 2020, https://mncfn.ca/the-toronto-purchase -treaty-no-13-1805/#:~:text=On%20August%201%2C%201805%2C%20 the,the%201805%20Toronto%20Purchase%20Treaty.

66 Mike Layton (city councillor), interview with the author, 6 February 2020.

67 Corbiere, interview with author.

68 Jon Johnson (University of Toronto), interview with author, 6 February 2020.

69 Clara MacCallum Fraser (Shared Path Consultation Initiative), interview with author, 6 January 2020.

70 MacCallum Fraser, interview with author.

71 David Delgado Shorter, "Spirituality," in *The Oxford Handbook of American Indian History*, ed. Frederick E. Hoxie (New York: Oxford University Press, 2016), 433.

72 Cote, interview with author.

73 Arturo Escobar, "Thinking-Feeling with the Earth: Territorial Struggles and the Ontological Dimension of the Epistemologies of the South," *AIBR Revista de Antropologia Iberoamericana* 11, no. 1 (2016): 11–32.

74 Indigenous Affair Office staff member, interview with author, 16 January 2020.

75 Layton, interview with author.

76 "Archived – Provincial Policy Statement, 2014," Ontario Ministry of Municipal Affairs and Housing, 5 November 2018, https://www.ontario.ca /document/provincial-policy-statement-2014.

77 MacCallum Fraser, interview with author.

78 Bob Joseph, "What Is the Seventh Generation Principle?," Indigenous Corporate Training Inc., 30 May 2020, https://www.ictinc.ca/blog /seventh-generation-principle.

79 Indigenous Affair Office staff member, interview with author; Indigenous Planning Perspective Trask Force, *Indigenous Perspectives in Planning*.

80 Indigenous Affair Office staff member, interview with author.

81 Corbiere, interview with author.
82 Matunga, "Theorizing Indigenous Planning," 31.
83 Borrows, "Living between Water and Rocks."
84 Escobar, "Thinking-Feeling with the Earth."
85 Deborah McGregor, "Traditional Knowledge: Considerations for Protecting Water in Ontario," *International Indigenous Policy Journal* 3, no. 3 (2012): 3.
86 Corbiere, interview with author.
87 Cote, interview with author.

Chapter Five

"My Story"

WADE HOULE

I wrote and submitted this essay because I finally gained the confidence to do so. So much about writing is about sending it out into the world. It feels like a tremendous risk, especially because the topic of the essay is me. It is a very personal piece, a small but integral part of history, my history. I wrote this piece because Indigenous history has often been ignored or omitted, and I wanted my children to have something that would stay with them forever.

Wasii'aa Giizhigo Inini

There is a boy who stands on the shore of a lake. He loves the water, and the sky. He is also very shy and talks only when he has to. He talks only when it is needed. He is always listening and learning, and he loves a good story.

On this particular day, the sky is blue, and it seems like it goes on forever. He throws a rock. He is looking at the endless sky and watching the calm, rippling water. He watches the tiny waves and ripples that move both away and towards him. Some of those ripples reach his toes, and some of those ripples go farther and deeper out towards the water; yet he is connected to them all.

He reaches down and touches the water and hopes that it helps him find the answers he seeks. He then raises his hands to the sky, with the water dripping from his fingertips. He does this and prays.

Why Am I Here?

When I was twelve years old I travelled from Dauphin, Manitoba, to the Long Plains First Nation, near Portage la Prairie, with my mother. It was a quiet ride, as per usual when travelling with my mother. We are both reflective people, and I do not remember, besides your normal chit-chat conversation, anything especially particular about what we said that day. It was near the end of winter when the south wind starts to bring warmer air into Manitoba and the snow starts to look grey and dirty; always an exciting time of the year.

We pulled up to a regular-looking house, much like many reservation homes, which had a centre front door and a centre back door. It had a small six-by-six deck to enter the door in the front, and the first thing you saw from the entryway was a living room to the right. The only difference in this particular home is that it was decorated with some of the most beautiful pieces of Indigenous artisanship I have ever seen. It was like walking into a museum. My mother and I were greeted by a woman, and they spoke Anishinaabe to each other. We laid our coats on the sofa and elder Don Daniels walked out from the back kitchen.

There was always something special about Don. He had an aura about him, and he always spoke with such grace, humility, and kindness. We shook hands and we sat down. As is customary in my family, I didn't say much because the adults were talking. I was never told that this was our custom, but it was always something my brother and I did; if the adults in the room are speaking, then you are expected to listen and be respectful. I sat there in awe both listening and observing. They spoke in Anishinaabe, and my eyes wandered the room. There were beautiful paintings, extraordinary sculptures, crafts made of antler and wood, and some of the most extravagant dreamcatchers I had ever seen. Don was not a flashy person, but these artefacts represented the amount of sacrifice, respect, and honour that was bestowed upon him. I knew that he was a special person, and many other people thought the same thing and honoured him as such.

In Anishinaabe culture, Don was considered a medicine man. Whenever my parents spoke of medicine people, it was with respect because they were gifted people. Medicine people have a tremendous amount of responsibility, and sacrifice much of their lives for the greater good of everybody. Often, the biggest sacrifice made by medicine people is time. Time is taken away from their families, time is placed in the energies of other people, and this takes them away from their spouses, children, and grandchildren. They give the ultimate sacrifice and are selfless.

On this particular day, Don was taking time to spend with me. My mother had brought me to his home so I could receive my Spirit name. In my family, we often refer to these names as "Indian names." I understand this is not politically correct, but as my parents usually say in reference to older slang or Anishinaabe translations, "I don't know, it's just what people call it." Hereinafter, I will refer to it only as a Spirit name. My mother pulled tobacco out of her purse and handed it to Don. Tobacco, one of the sacred medicines, is a customary exchange for the knowledge and wisdom that people have. It is an act of humility, respect, and love. There is always a moment, a split second, of energy that is shared when these exchanges occur. In the gifting of tobacco, it allows the people involved to centre themselves, and it provides purpose and reason.

Don led us down the hallway of his home. On the first door to the right was the bathroom; we took the first turn to the left into a small bedroom. Before entering the room I could see that there were two more bedroom doors a few steps down the hall. The design of this home was familiar and much like my uncle's home. I felt very comfortable and warm. We entered the bedroom, which once again was adorned with Indigenous artefacts that I could only assume were gifted to elder Don. The room was dark, and my mother and I sat on the floor beside a bed and Don sat on the floor preparing for the naming ceremony.

Knowing your Spirit name can be integral to an Indigenous person's identity. It provides purpose and direction, and in "Naming" below, I share the story of how my two daughters, Grace and Natalie, came to receive their names.

Naming

There were two beautiful little girls sitting in a darkened living room on a cool, fall afternoon. With them were their parents, grandparents, and two elders.

The elders prayed and prayed, and they sang song after song. Soon after, the woman elder spoke and she said, "This little girl right here will be known as Gabiidaabang Ikwe, Rising Sun Woman. She will bring tremendous energy and life to the lives of people every day." The wolf, ma'ingaan, will be her guide.

The elder continued to speak: "And, her big sister will be known as Misko Miki-nakens Ikwe, Little Red Turtle Woman. And, she has a gift that she will share with the world. That gift is in a spider's web in her hands, and she will have many dreams and heal many people." The turtle, mikinak, will be her guide.

This is who these girls are, and this is what these girls will be. In their lives they will learn to speak Ojibwe and be Anishinaabekwek, Ojibwe women. They must honour this every spring, and every fall.

After Canada's creation of the *Indian Act* in 1876, these types of ceremonies were banned and outlawed and many people practised in private or in hiding.[1] Ceremonies, sweat lodges, shaking tents, medicine bundles, powwows, and celebrations were all outlawed. The *Indian Act* also circumvented the Numbered Treaties and allowed chief superintendents to create the reservation allotments and the Canadian residential school system. Being born when I was, I was not directly affected by these policies, but I have inherited the colonial effects of these paternalistic decisions. It is an odd feeling to feel both relieved and privileged to not have to go through some of these acts of legislation, but to also feel the rightful pain, anger, and hurt for my ancestors who did.

In 2008, the Truth and Reconciliation Commission of Canada (hereinafter referred to as the TRC) was established to honour the stories of residential school survivors. In 2015, the Government of Canada called on the citizens of our country "to redress the legacy of residential schools and advance the process of Canadian reconciliation"[2] with Indigenous peoples. The TRC released 94 Calls to Action. The chief commissioner of the TRC is Senator Murray Sinclair, who is also a former honourable justice of Manitoba's Court of Queen's Bench.

In a speech to delegates at the Mosaic Institute Peace Patron Dinner in 2016, Senator Sinclair addresses the crowd about the identity, and the future identity, of Canada's Indigenous population, especially young people. Having a sense of identity is integral to any culture and any person's self-worth, self-esteem, and purpose. In his speech to delegates, Sinclair says,

> It boils down to four very important questions: you have to know *where you come from* … It is also about what is our creation story, the history of your people … If you know the answer to that question, then you will be able to help it yourself answer the next question, which is *where are you going?* … All of that is about faith, a sense of hope, a sense of future … We also have to answer that third question, which is *why am I here?* What is our purpose in life? … If you know the answers to those three questions, then you can answer the fourth question for yourself and that is *who am I?* Who am I is *the* question … the one that we are always challenging ourselves to be, the one that we are always trying to figure out.[3]

Senator Sinclair poses these four questions to Canadians: Where do I come from? Where am I going? Why am I here? And who am I? These questions are essential to identity. They are very difficult to answer, but people must attempt to do so in order to understand and situate ourselves in the fabric of our society and attempt to understand our shared history. In doing this work, I am attempting to do just that. I am attempting to figure out who I am.

Sinclair is providing the Canadian public a fundamental teaching in Anishinaabe culture. The questions posed might be new and novel for some, but these are questions I have heard my entire life. Since I was

child, and into adulthood, I have consistently been asked these questions. They surfaced in everyday conversations with teachers from my
early schooling in Indigenous communities, in interactions with elders,
and when meeting Indigenous people in order to situate ourselves via
our communities and our family members. We are consistently referencing these questions in everyday life, and although the answers we seek
are not always there, we understand the questions remain the same our
entire existence. And sometimes, we find our reasons through the Spirit
naming ceremony.

That day in Don's small but warm home, and in that small and darkened room, I received my purpose. I was gifted and honoured with my
Spirit name. Much of the ceremony of my name was done in Anishinaabe, and the majority of what happened that afternoon had to be
translated by my mother. I understand very little of our traditional language and I always have to refer and defer to my parents when it comes
to Anishinaabemowin. I was told that afternoon my name, which already
existed in me when I was born. After the ceremony, Don and my mother
exchanged words again and he explained to me in English where my
name came from. He said that I was placed on the earth to bring bright-
ness to people's lives. And that, when the sky is at its bluest, the ancestral
relatives and the Creator were communicating and reminding me of my
purpose. It was an amazing feeling that connected me to my history, my
people, the land, and my culture. It is my Spirit name that has guided
me ever since. In that moment, like Sinclair states, it was "the name of
the spirit that was placed in you when you were created, by the Creator."[4]
It is this name that has helped me find my purpose, and reflecting back
on this, it has shown me that purpose starts with something as simple as
a name.

Names are important, especially for Indigenous peoples. We live in
a settler society, a Western society, that is not always welcoming to our
beliefs, our traditions, our truths, and our names. So at times we can
feel unsafe in sharing something so intimate. And regrettably, when we
do tell non-Indigenous people our names, there is always a chance that
it is followed by ridicule, or sarcasm, references to stereotypical or silly
names from TV or movies, or questions about culture that are often tir-
ing and exhausting to answer time after time. To share a name requires
trust and a relationship strong enough that you do not feel judged.
When trust is established, then the listener comes to understand that
Indigenous people's Spirit names are symbolic of many things, but they
ultimately mean that we walk in two worlds: Indigenous and Western.
We exist in two places, and that is not always fully understood; therefore,
Spirit names are not always shared.

I do not remember the drive home afterwards. All I remember is that I was with my mother. My mother, Margaret, is shy and quiet. She enjoys laughing and staying connected with her family via stories, phone calls, and watching grandchildren. She has been a schoolteacher for over forty years and has always kept me grounded and connected to my culture. My father, Russell, worked for the railroad, travelled lots, and has never been a firm believer in the spiritual ways of the Anishinaabe people. Interestingly enough, his first language is Anishinaabemowin, and he grew up surrounded by the culture. Yet, he has never truly been a believer in the traditional customs and ceremonies of our people. In no way does he mock or disrespect these beliefs and traditions, but he is not a consistent practitioner of Anishinaabe ways.

For my entire life, my mother has always been there. I rarely remember a time in my childhood that I was away from my mom. My father travelled with his job, and he often left for work in the early years of my life, so it was my mother, and grandmother, who looked after my brother and me. My brother Kevin and I were active boys who loved to play hockey and baseball. Therefore, my mother was our driver. My older sister Elaine and my brother Kelly were close to adulthood and had moved on from our home when I was young. Elaine went off to school and worked in Winnipeg. She ultimately became a teacher and works in Ebb and Flow with my mother. Kelly started a family young, and he left his schooling to go and work and support his family. He also moved back and works within the Ebb and Flow community.

All of this made my brother Kevin and I very close, and we spent a lot of time together as children. My mother would drive us to our hockey games in the winter, and she would drive us to our baseball games in the early summer. If we were not playing sports, then she would drop us off at our grandparents' home for the summer holidays. My father is extremely reliable, and I depend on him for so much, including advice that has led me down my current path, but whenever it came to traditional teachings or Anishinaabe spiritual ways, it is my mother whom I would talk to. She always had a reflective answer to my questions, and if she didn't, she would seek out those answers from the people she trusted.

I am forever grateful for my mom and her spiritual guidance. It was her decision to take me to Don Daniel's home, knowing full well that going through that sacred naming ceremony would provide guidance and assurance in my life as an adult. I understand that it was the grandmothers and grandfathers of my ancestral line, along with the Creator, that placed my Spirit name in me, but it is my mother who showed me how to go through the process so that I could understand and have that knowledge and experience to pass on to my own children. It is because of her that I know *why I am here*. Life is a journey, and so is this work, and it is the building of a trustful relationship between researcher and reader that allows me to say that my name is Wasii'aa Giizhigo Inini, Bright Sky Man.

East

Nanshee always loved the springtime. Spring, just like the sun, always brings new beginnings.

Well, one day, she woke up before the sun rose and made a warm fire in the stove.

She then placed a kettle on the stove and warmed the water for her morning tea. She prepared all her seeds and plants as her tea boiled and then cooled.

When her tea was ready she took a few sips to start her day. As the sun started to rise in the East she tied her scarf under her chin so that it covered her head. She was heading out to the field, near the lake, where she was going to plant her garden. She took her hoe and her shovel. She hammered in sticks on both ends of the garden and tied a string all the way across them to make a straight line. Then, she planted her seeds and softly covered them with dirt. Her garden was going to be huge.

She never used gloves. What for? Your hands should get dirty and feel the earth.

Where Do I Come From?

I remember my great grandmother Nancy. I was small, maybe seven or eight years old, when I would go and visit her. My kookoo Nancy lived with her son Ernest, my great uncle. She was born in 1899, and she was small but wore the experience of perseverance and resiliency on her skin and in her eyes. She only spoke Anishinaabe. I never verbally communicated with her, but she had this sly and warm smile and a genuine look in her eye when I knew the adults were speaking about me, or if she asked me to do something for her.

Her home was small, and she never needed the modern comforts of home that many people desire these days. She didn't watch TV much, but when she did it was the soap opera *All My Children*. My uncle Ernest would translate for her the English dialogue that was happening on the screen. She had a small bedroom she slept in, but she also had a small bed/couch in the corner that she would sit at, if she wasn't sitting at the kitchen table. There was a stove near the entrance, and it was a comforting place to be. As was a regular occurrence in those days, she often wore a handkerchief around her head with it tied around the bottom of her chin. If there was ever a picture in the dictionary of an elder Ojibwe kookoo, her face with that handkerchief on would be it.

Nancy was born in Fairford, Manitoba, and she married my great-grandfather Charles Maytwayashing in 1921. Charles was from the Lake Manitoba First Nation, and he died in the Selkirk Mental Hospital in 1971. I include this piece of information because it is important. It is important because my mother, Margaret, states that the odds were that Charles had what we would now know as Alzheimer's disease. He started to lose his memory early in life and would do things like lose direction, misplace things, and tell lies to make up for his forgetfulness. In those days people just assumed doctors were right because they were doctors, but when their daughter Ida (my grandmother) got Alzheimer's in the 1990s, pieces to Charles's story were put together, and we came to understand that this disease was passed down hereditarily from him to her.

People in the community called my kookoo Nancy: "Nanshee." She and Charles had a log cabin along the shores of Lake Manitoba where they had a huge garden. My kookoo Nancy loved to garden. She also loved to tell stories. She was a storyteller. And, often, as is common with storytellers, she liked to embellish stories about people in the community and gossip about their exploits. I was never privy to these stories because I could never fully understand Anishinaabemowin. That being said, she wore expressions and experiences on her face like granite on the shores of a lake. Her eyes were piercing, and I would watch words form on her lips. I would listen and hear, but it was her eyes and hands that I watched. I could feel them. She would weave humour, embellishment, love, and a crafty and sly sense into her everyday life. I was fascinated by her.

Nancy truly cared for people. Whenever someone died in the community, she would pack her things in an overnight bag and travel to the home of the people who had lost their loved one. She knew the hurt those people were going through; she helped to look after children and she would do all the cooking and cleaning and prepare meals for the wake service and the visitors who would stop by the homes to pay their respects. She understood people's pain, and Sherman Alexie, in his book *The Absolutely True Diary of a Part-Time Indian*, talks about funerals and death and how it is related to crying and laughter. He states, "when it comes to death, we know that laughter and tears are pretty much the same thing ... and when we said good-bye to one [person], we said good-bye to all of them. Each funeral was a funeral for all of us."[5] It is what Alexie is explaining here that Nancy fully understood and that made her simply incredible and selfless.

In discovering where I come from, I know that her story is integral to mine. She is symbolic in her perseverance and resiliency of the women in my ancestral lineage because, although she was small in frame, she was big in heart and voice. She died in 1993, and I miss her. I wish I could go back and spend more of that quiet time with her – those small moments we shared when no words were spoken.

Born of Nancy and Charles' relationship was my grandmother Ida. My kookoo Ida was an amazing person, and I spent a lot of time with her. My mother would drive us out to my grandparents' home in Vogar on the weekends or for our summer holidays.

My kookoo Ida was born in 1933 and married my grandfather Abraham in 1951. My grandfather Abraham built a log house on the shores of Centre Lake, which is near the small, quaint village of Vogar, Manitoba. As their family grew with the arrival of my mother, uncles, and aunt, they would eventually move into the village of Vogar, then off to Eriksdale to be closer to the farm my grandfather worked at, and lastly, moving back to Vogar in 1980.

It is this last home in Vogar that I remember fondly. The home I spent so many days and nights at is what I consider home. I loved going there. Coincidentally, their new home was actually about a mile from their original home near Centre Lake. As a child, I would often walk to Centre Lake because I was playing in the bush and exploring the forests near their home. I was always outside during the day, and at night I would either play or watch hockey or baseball with my grandfather. As I got older my brother or my cousins and I would quad or snowmobile down to Centre Lake.

My kookoo Ida always had delicious food. It was simple and humble food, and I was always accustomed to a big pot of soup on the stove simmering all afternoon. My grandfather Abraham loved to visit and talk about old times, so he always had a lot of visitors. As was customary in those days, reflecting the way that generation was raised, there was always food available at anyone's house at any given time. Soups as simple as garden tomatoes and macaroni, or hamburger and macaroni, or duck or rabbit soup were often simmering on my kookoo's stove, ready for her next visitor or hungry grandchild. When you walked into their small kitchen, there would always be bannock on the right-hand counter. It was always best when it was warm, and I always knew if there was fresh bannock because you could smell it everywhere, or it would be resting sideways up against the flour and sugar containers because that was the way she cooled it off. I can still taste that warm bannock and the butter melting on it, ready to eat with that fresh hamburger and macaroni soup. Anytime I have that combination to this day, I am taken back to those warm memories of my kookoo.

Ida was a caregiver. Much like her mother, Nancy, she looked after people. She always had food ready for visitors and she welcomed people into her home to spend the night. My brother Kevin was born in 1975, when my mother was in the midst of her teacher training. My kookoo kept Kevin in Eriksdale while my mother went to Brandon University, and when she got her first teaching job in Jackhead First Nation my kookoo once again kept Kevin in those formative years. She cared for people and loved her grandchildren dearly.

In the 1960s, when my mother was a teenager, my kookoo Ida would take my mom to Winnipeg to help look after relatives. Ida's niece, my mother's cousin, was struggling with alcohol addiction, an abusive relationship, and sang in a music band on the weekends. The weekend gigs and binge drinking were hard on the young children, and my kookoo would travel with my mother to look after those kids. Unfortunately, those children would ultimately be taken by Child Services and my mother's cousin, Hazel, would never see them again. Later in life, Hazel would succumb to alcoholism, and she never did see her children because they became a part of what is now known as the Sixties Scoop. Those children were taken and needed to be in protection, but it was never in the plans of our family to not see them again. There was no consultation or parent plan, there was no treatment or rehabilitation programming, and zero effort was made to offer therapy to Hazel and help reunite her with her family. There was never an intention to keep them together, or at least with other family relatives or in the same community.

The Sixties Scoop in fact took place from the 1960s to the 1980s, at the time the Canadian Indian residential school system was being phased out, where government officials purposely removed Indigenous children and placed them in foster homes all over the world. In 1998, one of those children contacted my mother via mail to initiate a relationship in order to (re)discover his family roots. Sadly, Lawrence was in a penitentiary in the state of New York and was looking for his siblings. This process started to unfold in the years following that first initial contact, and we were able to meet and reunite with seven of the nine brothers and sisters of the family. It was an exhilarating and exciting time as we were meeting long-lost family members, but it was also a tough time for those siblings because they had been apart for so long. Upon discovering the whereabouts of some of the children, some were moved to Pennsylvania, North Dakota, South Dakota, and some were placed right within Winnipeg. Sadly, two remain out in the world somewhere.

This type of event would be traumatizing to anyone, and like I stated earlier, Hazel never recovered from her addiction – indeed it only intensified. This colonial approach to dealing with Indigenous people was traumatizing for my grandmother and mother as well, but it was not uncommon. Colonialism and Canada's history and relationship with Indigenous people has been tumultuous. Policies such as the *Gradual Civilization Act* of 1857, which was created ten years prior to Canada becoming a country, was designed to assimilate Indigenous people. It was a paternalistic document that encouraged assimilation and called for "Indians in good moral standing" to adhere to the ideals of European land ownership.[6] Colonial attitudes and beliefs like this were forced upon Indigenous peoples. Moving towards the creation of our country in 1867, dealing with Indigenous peoples was paramount. The *Indian Act* of 1876 is evidence of this assimilative practice because it superseded the Numbered Treaties, which were signed on the Prairies starting in 1871, and it allowed the Canadian government to circumvent its constitutional obligations. In 1885 and through to the 1940s (it was officially repealed in 1951), the government also implemented the pass and permit system, which required First Nation farmers to obtain a permit to legally sell their products off-reserve. The Indian agent, which controlled First Nation communities, controlled and distributed the permits. The system was restrictive and Indian agents would often not grant permits, which would leave crops and produce to rot in the fields. To go with this practice, passes were also handed out to residents living on reservations where the Indian agent would allow people to leave the community to go to nearby urban centres, visit family in neighbouring communities, or to seek employment. As was the norm, time limits were often placed on people, and they could be arrested if they ignored these limits, and if they took too long or obtained jobs outside of the community then the members could be taken off the band register. This is an important history that often goes untold in Canadian educational institutions.

Policies such as the *Gradual Civilization Act*, the *Indian Act*, the pass and permit system, enfranchisement, Indian residential schools, the Sixties Scoop, and the large number of children currently in the foster care system can arguably be attributed to the paternalistic attitude and discriminatory legislation that has been forcibly imposed on Indigenous people in Canada.[7] I understand that this legislated discrimination is important to who I am, but it is not my focus. I cannot move through this essay without mentioning the types of horrors Indigenous people faced, but I do not want it to be what I concentrate my writing on. Chimamanda Adichie, in her popular TED Talk entitled "The Danger of a Single Story," states, "all these stories make me who I am. But to insist on only these negative stories is to flatten my experience and overlook the many other stories that formed me."[8] The resiliency and perseverance that all Indigenous peoples have, just like my grandmothers, are especially important stories that need to be told. They need to be honoured, and I wish to do so by "*re*writing and *re*righting our position in history."[9]

My kookoo Ida was a band member of the Lake Manitoba First Nation her entire life. As a child, she lived on the shores of Lake Manitoba itself, which was on reserve land. When she married my grandfather in 1951 at the age of eighteen, she lost her status as an Indian under the *Indian Act.* This meant that she was not allowed to live on the reserve, near her mother, and she ultimately moved to Vogar, the nearby Métis community. My grandfather Abraham was considered Métis and grew up near the community of St. Laurent, one of the largest Métis communities in Canada. "The paternalistic manner in which [Canadian] governments manage the affairs of Native people"[10] led to the creation of the *Indian Act* in 1876. The *Indian Act* and its enfranchisement rules stripped women of their Indian status once they married a non-status man. It only allowed men to retain their status as Indians when they married non-status women, and interestingly enough, non-status women (e.g., German, Ukrainian, Chinese, etc.) gained status rights under the *Indian Act.* In those days, once Indian status was lost, you were unable to live on a First Nation reservation in Canada. This forced assimilation and enfranchisement was common practice.[11]

Despite the forced effects of colonialism and the attitudes of settlers in Canada, my kookoo Ida persevered. She was tough. She was tough mentally and physically. When I was a child, and as she neared retirement, she worked at a small resort called the Narrows Lodge. It wasn't far from Vogar, and she would work at the restaurant and help to clean the hotel rooms at the lodge. I remember going there and running around and playing. Although she was pretty shy, she had a great sense of humour. I recall fondly her eyes and her smile. She liked to laugh, and because my grandfather was funny and crazy at the same time, she laughed all the time. When she gave her biggest laughs, she would bring one hand to her mouth to cover it and raise her head in the air to try and contain it. When I was five, I walked out from their kitchen into the living room, where she was sitting on the recliner, and I said, "Kookoo, how come the chicken couldn't cross the road?" She said, "I don't know, how come?" I replied, "Because, it had no legs." I promptly walked away after telling her the joke and she started to laugh. She couldn't stop laughing; and, just when I thought she was done laughing, she started to laugh again. I'll never forget that moment. It often took her about twenty minutes to stop laughing.

When I was about ten years old I started to notice little quirks in my kookoo. One day, she was sitting at the end of the couch like she always did. She had her glasses on, and she really wasn't doing anything but looking at the TV or glancing out the front living room window. She put her glasses down, got up and started walking towards the kitchen. She took about four steps, stopped and looked around, and asked where her glasses were. I thought she was joking, but there was a genuine look on her face and I could tell she was serious. When I told her where they were, she showed a small sense of embarrassment, but also a stern look of concern. What I did not realize at the time was that she was slowly starting to lose her memory and that Alzheimer's disease was starting to set in. It was slow at first, but those types of incidents started to become more frequent and soon she was leaving stoves on, forgetting to eat, and missing steps walking up and down stairs. It was becoming very dangerous, and one day she fell. It was a serious fall, and one that she never really recovered from; afterwards, her disease progressed more rapidly. She was soon confined to a wheelchair with an in-home care service coming more consistently, especially when my grandfather started to struggle to look after her.

Honestly, it was awful to witness. I felt like I never got a chance to say goodbye to her. She was in the room, but she was not present. There were times, though, when a little bit of her would come through. I was on the couch one day lying down and she was seated in her wheelchair near my feet. I must have had a small hole in the bottom of my sock because she reached out and stuck her finger in and wiggled it around. She tickled me. That was the last interaction I had with my kookoo where I knew she was in the room with me. I consciously think about her every day in the hopes that she follows me from the spirit world.

Like my kookoo Nancy, I miss my kookoo Ida dearly.

Being Young

When Margaret was a young girl she was always playing outside. That is what everybody did in those days, and that is what people should do now. But you know, people sometimes forget their ways. It is both funny and sad how people can lose things so quickly.

Anyways, when Margaret was outside, she was always playing with her cousins. Playing with cousins is always exciting. It is just so fun to be young, to be small, to laugh, to cry, to breathe heavy from running, to scream in joy and excitement, and to simply just sit and visit and talk.

Sitting and talking and doing nothing can be fun. It is fun because you have to get creative. You have to get up and pretend and create an imaginary world that you can do anything in.

This one time, Margaret and her cousins found some old tires in the bush. They pulled them out of the weeds, cleaned them up, poured the water out of the insides, and started rolling them. Those tires rolled so fast that they pretended they were cars. They ran and ran, and they pushed and pushed those tires, screaming and yelling and laughing the whole time.

My mother Margaret grew up in Vogar, Manitoba. This small Métis community is near the shores of three lakes: Centre Lake, Dog Lake, and Lake Manitoba. It is also the neighbouring community of Lake Manitoba First Nation. Vogar is a small village. As a young child, my mother's father, Abraham, my grandfather, built a one-room log cabin for his family to live in. It was a humble home that my mother remembers fondly. Like any home it was warm and loving. And, as was common in those days, it was close to nature. You were either in the house, or outside.

The house was small with a single fireplace, two beds, and a kitchen table and stove in the corner. My grandparents had one bed, and my mother had the other bed. When someone came to visit, often they would spend the night, which relegated my mother either to the floor or to my grandparents' bed. This was common and normal.

Not long after, as my grandparents started a young family, they would move into the village into a bigger home and were located closer to other families, and closer to school. It was nice for my mother to be closer to the community at that age, and she loved playing with her cousins who lived nearby. My grandfather got a job working on a farm near Eriksdale, Manitoba, which was about fifty kilometres away from Vogar. Again, my grandparents moved the family to a new home to be closer to work. My grandparents decided to move to Eriksdale and raised my mother and her three siblings there.

My mother describes life as simple, but people worked hard. And she really emphasizes the fact that people worked. That has always been important in my family, and that value has carried on throughout the years. My grandfather, my mom's dad, was adamant that you worked. You either worked or went to school. Those were the only two options in life, and my mother inherited those values and have passed them down to me.

My mother loves Vogar. She talks about it fondly, and very often. I stated earlier that my mother grew up in a small home, moved to another small home, and eventually moved to Eriksdale. Eriksdale was a small town filled with settler farm families. My mother made good friends in Eriksdale, but she does not speak as lovingly about her experiences in the school. Schools are not designed for Indigenous children,[12] and often she was the only Indigenous child in her classroom. She had relatives who would be bussed in from Lake Manitoba First Nation, but there were many times when those kids did not make the bus and thus they did not show up to school. In Eriksdale, at school, my mother would wait for the buses to arrive. If she did not see any familiar faces come off the bus she said that "would make for a long day." My mother implies here that without the comfort of her people (Indigenous culture) and family, she felt she did not belong. What she does not say here is that what she feels is racism. It is not the overt racism that we often associate with images of cape- and mask-wearing Ku Klux Klansmen or horrific images of Black people being attacked, or mauled by dogs, or having hoses sprayed at them during the civil rights movement in the United States, but it is the racism that is intangible and difficult to identify. Robin DiAngelo says that "racism is a system" and it "does not refer to individual white people and their individual intentions or actions but to an overarching political, economic, and social system of domination."[13] Simply put, because my mother was Anishinaabe, she felt she did not belong, and racism was the main contributing factor to this lack of connection.

My mother is smart. She is quiet and humble, and she has a calm demeanour that is comforting and welcoming. She excelled in school and graduated from Eriksdale in 1970. She always wanted to be a flight attendant or airline hostess. She wanted to travel the world, and she loved the way the women looked and the way they dressed. It seemed glamorous. As high school was nearing an end, she started to volunteer at the local hospital as a candy striper, and this eventually turned into a job. She learned that she worked well with people. Something was missing, though, so she moved back to Vogar to work at the school there as an educational assistant. A superintendent of the school division was visiting one day and asked my mother if she was interested in becoming a teacher. The Government of Manitoba was offering funding for Indian and Métis people to get certified as teachers. She knew full well that neither herself nor her parents were able to afford such an opportunity, so my mother jumped on it. She registered and hopped on a bus to Brandon University to start her work in education. It wasn't exactly what she was interested in, but she grew to love it. My mother is still a teacher to this day, and in October of 2018, at the age of sixty-six, received her post-baccalaureate in special education from the University of Manitoba. An incredible feat, considering she has been a teacher for over forty years and still took the time to study while holding a full-time teaching position on the Ebb and Flow First Nation.

It was at Brandon University where my mother studied to be a teacher in the beginning. While studying to be a teacher she had my brother Kevin, along with meeting my father, Russell. As my mother neared the end of her teacher training, my kookoo Ida looked after Kevin and my father travelled in north-western Ontario on the gangs of the Canadian

National Railroad (CNR). Having children myself, I cannot imagine having to leave my child in order to secure a position in my career of choice. That is exactly what my mother did. That being said, my kookoo Ida was perfect for the job. Much like my own mother, she loved children. My mother made major sacrifices for my brother, and I would soon come along in 1980 while my mother taught on the Fisher River Cree Nation. A few years after I was born, my mother would move back to Vogar once again. This time she came back as a teacher. We lived right near the school in a teacherage, and we were able to walk across a short field to get to class. Near our house and the school was an outdoor hockey rink with a wood-stove-heated shack. My brother and I spent a lot of time on that rink, although I was rarely allowed to play late because the big boys were out there, and I was too small.

I went to kindergarten at Vogar School and I loved it. It was a small school and Mrs. Johnson was our teacher. There may have been five or six of us on any given day. I would soon move to grade 1. My mother was my grade 1 teacher. But she was also the teacher of other students all the way up to grade 6. There were about fourteen of us ranging from grades 1 to 6, and my mother was the lone teacher. I was young and small, and my memories are scarce, but I loved that school. It had a short hallway that led to the gymnasium, and we spent a lot of time playing and running around on that dark-green gym floor. I am sure we learned as well.

The school would eventually close after my grade 1 year, and in grade 2 I started to bus to the nearest town, some fifty kilometres away in Ashern, Manitoba. This was around 1986, and at that time my father would purchase a house in Dauphin, where he secured a permanent position with the CNR in the town of seven thousand. My mother taught at Ashern Elementary while I was in grade 2, and in 1988 she got a position on the Peguis First Nation. It is in Peguis where I was truly introduced to hockey. Peguis lives and breathes hockey. I really had no choice because all my friends played, so my mother signed me up too. It was the first time I played organized hockey, and I wasn't just playing shinny with my brother on the outdoor rink. My brother and I spent many nights at the outdoor rink or at public skating, where he taught me to shoot, pass, and whip around the ice. If I wasn't playing hockey, then my mother, brother, and I would travel to Dauphin to spend the weekends with my dad. As weird as their relationship may sound at this point, they were in fact together. My mom worked and lived in Peguis during the week, and then we either went to Dauphin or travelled for hockey games on the weekends.

My parents, especially my mother, were selfless in their approach to our rearing. They led by example and made sacrifices that ensured that my brother and I would become successful adults. Despite many obstacles in my mother's life, my kookoo's life, and my kookoo Nancy's life, these women lead by example with a humble and courageous approach to life. I have absorbed these values and ways of being, and I nervously and humbly tell the reader that I am attempting to follow in these women's moccasins. More importantly, I think that I am supposed to do this.

My mother is the pillar of our family. She is the most consistent person I have had in my life, and she has always been by my side. Come springtime, my mother would sign my brother and I up for baseball. However, baseball wasn't in Peguis, it was in Dauphin, and we were still living in Peguis. When the snow melted we would play baseball, travel to Dauphin, and stay with my father. Growing up in Vogar meant that everybody played baseball. Baseball was another passion in our family. Playing these sports, and eventually becoming competitive players, gave my brother and I the opportunity to develop leadership qualities. Sports helped us to develop confidence, teamwork, and communication skills that we both employ now in our positions as public servants. My brother works for Manitoba Hydro in accounting, and I am a teacher.

We lived in Peguis for five years, and we all eventually moved in with my father, in Dauphin, in my grade 8 school year. It was 1993 and it was the first time that we all truly lived together, every single day. Travelling, going in and out from my mother's home to my father's home, and also to my grandparents' home, was all normal to me as a kid. It was just simply how we operated,, and this is *where I come from.*

Where Am I Going?

Up until 1993 I had spent most of my childhood with Indigenous people. As diverse as Indigenous people are, the First Nations with whom I lived and spent most of my time were all people who had similar experiences as I did, they often looked the same as I did, and many of us understood that unspoken language of values, beliefs, and customs. I never questioned my identity up to this point. When I moved to Dauphin, I didn't have much choice but to question my Indigenous identity. It was pretty evident when I walked into my new classroom that I was one of few Indigenous people in the room.

Dauphin is a settler farming community and in the early 1990s was made up of mostly Ukrainian people. It's tough to be an Indigenous person in Dauphin. There is a total lack of understanding and education there, and it was pretty evident back then. I understand that I'm making a very sweeping statement here, but it is important for me to name the truth. Robin DiAngelo, in her bestselling book *White Fragility*, addresses similar sweeping statements: "the mere title of this book will cause resistance because I am breaking a cardinal rule of individualism – *I am generalizing* ... For now, try to let go of your individual narrative and grapple with the collective messages we all received as members of a larger shared culture [and] unsettle the racial quo."[14] Like I said, it was evident, and this lack of understanding was based on the underlying racial inequality that exists in Dauphin.

Dauphin sits on Treaty 2 territory. That agreement was signed in August of 1871. Not many people in the town of Dauphin to this day know that simple fact. Dauphin was also home to the Mackay Residential School. It was built in 1957 and closed in 1988; it was one of the last residential schools to close in that era. When I moved to Dauphin permanently in 1993, the residential school had actually been converted to a Christian college. It was called the Western Christian College, and they bought the property and school from the government in the early 1990s. I never thought much about it until I got to university in the early 2000s and learned in depth what residential schools were. All of those memories from 1986 and 1987, when my brother and I would go over there to skate and see all those Indigenous kids, started to come back to me. We even attended a graduation ceremony there one time, with pow-wow drummers and dancers. I didn't remember those memories until I learned what the actual building used to be. It's weird how that happens, when something triggers memories to come pouring in like that. It was in those university classrooms and lecture theatres that I came to realize that the building was a residential school, and it was a mere three blocks from my home in Dauphin. There was a sense of sadness as I came to learn and realize what may have happened, and probably happened, in that building. The pervasive abuse, violence, and humiliation that occurred in those buildings; but also the realization that the Canadian government and churches held so much power during that era, rendering so many children and families helpless. This knowledge humbled me, and I quickly realized the privilege I had.

I met a couple of neighbour kids back then; one was named John and the other was Tony. Tony's family was from the Atikokan area of northwestern Ontario, and they were extremely friendly and welcoming. Tony and I became close friends. Tony was much like me: we both loved to be outside and play in the bush. They were nice people. Tony's father worked on the railroad like my dad. We spent a lot of time together as youngsters. The family of the other boy, John, was not nearly as friendly. On a couple of occasions, John's family would not allow me into their home, and I never did see the inside of their place. I recall one time John's sister came out of their camper, which was pretty cool because in those days not a lot of people had campers. She looked directly at me, and then looked at her brother and said, "John, that boy is an Indian, and Indians are not allowed at our house." I was with Tony at the time, and we simply walked back down the block and went our separate ways. I told my mother later about what had happened, and she told me that I was never to go there again. And I never did.

Because I had spent parts of my summer playing baseball in Dauphin, the transition to my new school was not difficult and I made friends immediately. It was an exciting time in the fall, plus hockey was going to begin very soon. I definitely experienced some culture shock. I went to having almost all Indigenous teachers back in Peguis to there being no Indigenous teachers in my new school. I saw myself in no one besides my peers. In Indigenous communities, you have a common bond and a shared sense of humour related to cultural happenings, or people in the community. Everybody knows everybody, and it is that connection that helps Indigenous people relate to many different stories of themselves. Living in Dauphin changed in that way, and I only connected with my close friends, who were often teammates. I loved Dauphin but I did not have that natural feeling of being connected to everyone in the community. When I went to visit my white friends' homes, I was shocked and appalled by the way they spoke to their parents. They spoke to their parents in ways that I would never speak to mine. I thought it was disrespectful and their use of varying tones was extremely rude. It was certainly something that I would never get away with in my house.

Eventually, I would move on to high school and get comfortable with everyday life in Dauphin. In my grade 12 year, I moved to a small town in eastern Manitoba to play competitive hockey. It was a tough move, but my parents trusted me to make the decision to be billeted alone with a family there. When the hockey season ended, I moved back in the spring to complete my schooling in Dauphin. Upon my return to the school, I saw a girl.

Magic

One time, not so long ago, there was a little girl who lived on a mountain and was always talking. She would talk and talk and talk, and ask question after question after question. She did so much talking that she started singing.

Well guess what? She now sang and sang all the time!

Soon after she realized she could sing, she started to write down all kinds of ideas. Some of those ideas were good, but some of those ideas were bad. You see, she was not a bad person, but there were bad things going on in her life. Writing her ideas and singing her songs saved her from all those bad things.

She was magic. She used magic to write and write, and sing and sing. People always wanted to hear and see her sing. Her voice and songs were magic, and she used it to save herself, and to save other people.

I met my wife Desiree in 1998, my final year of high school, and we started dating in the summer. My wife is an incredible woman. She is a survivor. She is a survivor in so many ways and is symbolic of so many ideals in this work. She is resilient and was forced to grow up fast as a child. We met young, and she was strong-willed and strong-minded then, as she is today. She is tenacious yet kind, quick-witted and patient, and she is brilliant but humble. She holds me to account yet respects who I am as a person. I have grown, matured, and have become a strong Indigenous male because of her unabated support. I am who I am because of her.

My wife is the granddaughter of a residential school survivor. Needless to say, the horrors her family faced in Canada's residential school system have greatly affected her life. The trickle-down effect of the residential school system is inter-generational.[15] The trauma faced by children in the system is everlasting. Senator Sinclair, after he discussed and explained the four essential questions for reconciliation – Why am I here? Where do I come from? Where am I going? And who am I? – states, "For young Indigenous people who were taken away from their families and placed in residential school institutions, the ability to answer any of those questions was taken away from them by those schools."[16] What this means for me is that I have to observe and critically think how this is going to affect my own children. In one way or another, the residential school system is going to affect my family, especially my children. It will be up to them to reconcile with their own past along with Canada's colonial past. They will ask questions that some people may not be able to answer, comprehend, or fully understand. In doing this work, in writing my story, I am attempting to understand who I am in Canada's fabric. I am attempting to do this through love. It is my wish that they do the same.

Wab Kinew, in his book *The Reason You Walk*, sets out to reconcile his relationship with his father.[17] It is a beautiful journey of love, discovery, sacrifice, and understanding of a familial relationship that has been affected by colonialism, and ultimately the residential school system. Kinew's relationship with his father as a child was difficult and challenging. And, in his journey, he decides to repair this relationship and better understand his own lineage. He soon discovers that it was never his father's fault; rather, it was the residential school system. Kinew states, "We stand by those who came before us, hoping that those who come after us will honour us in the same way. We love, and we hope to be loved … so as long as anything other than love governs our relationship."[18]

As Kinew states, love governs our relationships, but my wife's grandmother and father were never loved. Their removal from their homes, their family connection, their language, the land, and their lifestyle altered them forever. I am not trying to assume that they were not loved at any time in their lives, or that they could not love others, but it is well known that committing cultural genocide and targeting children has altered the course of our country's narrative, and our people's knowledge of love. My wife's grandmother, and their relationship, has been shut off socially, emotionally, and physically. She has always struggled in showing love. When my wife's father was born, he was given up for adoption. He did not know this until he became an adult. So once again, that lack of love led him down a path of chaos and dysfunction. As a result, he struggled to show love and sought comfort and coped via irresponsible behaviour and abandoned my wife and her sister at a young age, along with their mother.

I would be remiss to do this work and not honour my wife and the obstacles she has been forced to overcome, but also to honour all the other children who are the survivors of residential school students, and the generations that passed, and that are coming. This work is important, a small but significant attempt at "redressing" those wrongs. Senator Sinclair says that "education got us into this mess; it will be education that gets us out."[19] It is my wish to honour my wife, her family's history, and my children who will one day come to understand and learn about their own family more closely.

Soon after my wife's father left, her mother met a tough Métis man. This new boyfriend worked as a diamond driller in the northern parts of Canada, but when he came home he partied hard and long and there were many nights my wife went to school on very little sleep and memories of yelling, laughing, and arguing long into the night. Maria Campbell, in her groundbreaking memoir *Halfbreed*, talks about the hardworking, but also hard-drinking, Métis men she grew up with in northern Saskatchewan. My wife's new stepfather was just that. Campbell says, "I never saw any of our men walk with their heads held high …However, when they were drunk they became aggressive and belligerent, [and] often they drank too much and became pathetic, sick men, crying about the past and fighting each other."[20] My wife and I met at a young age, so I was a personal witness to the behaviours of her stepfather, but like I said, my wife is a survivor. She has the heart, strength, and courage of a bear, and has shown that she can overcome anything.

All my wife ever wanted to do was to leave home and go to university. She found solace in education and excelled in learning. When she completed high school, we both signed up to go to university, and she attained a degree in law after a short stint in the Faculty of Social Work. She is extremely motivated and an amazing mother. Her motivation, hard work, and perseverance has helped her stop a cycle of dysfunction in her family. Education can do that.

That being said, my wife has fond memories of her childhood. Her own mother worked two jobs and supported the family during those difficult times after her father left. When her new stepfather entered the picture, there were still quiet periods during that time because he would often go off to work for long stretches, including months. It is those memories with her mom and her sister that she cherishes the most, and they grew closer during those times. In her childhood, like many young girls, my wife enjoyed music. She especially loved country music and grew up on classic country and western singers such as Dolly Parton, Loretta Lynn, and Reba McEntyre. At a young age my wife started writing songs. As many artists do, she wrote about the things she heard, saw, and experienced. Many of her songs are about alcoholism, partying, and tough times, but it is those songs that many people identified with. People relate to her music as alcoholism and addiction are universal and do not discriminate.

At the age of fourteen, my wife, Desiree, released her first independent studio album – an amazing feat considering the odds against her. Since recording that first album and going off to university and becoming a working professional, she has released three more albums and is currently working on her fifth. Simply incredible. She has been nominated for, and has won, numerous awards for recording and songwriting and was lucky enough to be nominated for a Juno Award in 2014. My wife, like my mother and grandmothers, is a strong and resilient Indigenous woman. All four of them have demonstrated that they are hardworking and humble, and that their past, and Canada's colonial past, does not define who they are.

Flower

I spent many nights waiting for momma in the car outside the local bar.
She said I'm just going in for a cold one, I knew that meant 6 more.
When she'd had her fill she'd take me home wasted behind the wheel.
I'd go to school the very next day like it was no big deal.
I'm a wait in the car kid while her mom's at the bar kid.
A riding home with a drunk kid, but I turned out alright.

It don't matter where you come from, cause it ain't where you're going.
Looking back, I grew up fast, just like a flower through a sidewalk crack.
We can't choose our mamas or determine our circumstance.
Raising babies, it ain't easy, you do the best that you can.
There was always food on the table and always love to go around.
We all make mistakes sometimes but somehow it all works out.
Now I'm a little older, got two daughters of my own.
I might stumble and I might fall, I was raised by a rolling stone.
One foot in front of the other, take it day by day.
I'm the furthest thing from perfect but you're never gonna hear them say:
I'm a wait in the car kid while her mom's at the bar kid.
A riding home with a drunk kid, but I turned out alright.
It don't matter where you come from, it ain't where you're going.
Looking back, I grew up fast, just like a flower through a sidewalk crack.[21]

– Desiree Dorion

Who Am I?

Like many people often do, I discovered a lot of who I am in university. It was fun and exhilarating and I met many people. University is also where I excelled academically. It was not that I was not a smart student, but I never put in a solid and concerted effort in high school. I was too focused on being a teenager, playing hockey, and was too self-absorbed.

I entered university the same time my wife did, and I was able to attain my bachelor of arts and my bachelor of education degrees. I was part of the University of Manitoba's Access Program, and they provided the necessary supports for both rural and Indigenous students attending post-secondary school. It was an amazing program, and if I did not have classes, then I spent a lot of my time at Access hanging out, using computers, talking to the different social and academic councillors there. It felt like home, and the people, atmosphere, and culture were extremely familiar. The program had a profound effect on me, and I still talk to some of the people involved in it today.

Along with sports, Access allowed me to be a leader and a role model. I was featured in their promotion and public relations posters and pamphlets, and they helped me get a part-time job speaking to schools and helping to recruit Indigenous high school students to sign up and register with the University of Manitoba. It was also while doing this part-time work, along with being one of only a few Indigenous education students in the faculty, that I noticed that I was having an impact on my peers and within my faculty.

I was becoming one of the role models of the program and people were looking to me to lead, speak on, and address Indigenous issues in education. I was consistently asked to sit on boards and committees, and although it was exciting, it also made me understand the lack of representation, the need for Indigenous peoples in all aspects of institutional education. This need for representation has never truly left, and I still witness it today.

Access prepared me for my first job interview in my final year in the Faculty of Education, and I was subsequently able to secure a position prior to graduating. I worked at Maples Collegiate in the Seven Oaks School Division for three years. My wife, in the meantime, quit social work and entered the Faculty of Law, and so I worked for a few years while she attained her degree. Maples Collegiate was a great school and I loved my time there, but living in the city was not for my wife and me. As soon as she graduated, we purchased a home in Dauphin, and I was hired on at the Dauphin Regional Comprehensive Secondary School. Like I said earlier, Dauphin can be tough place for an Indigenous person to live, but my wife and I felt it was important to move back, contribute to the community, and become leaders in our chosen fields. Since then, I have had the opportunity to sit on multiple Manitoba education committees and multiple Manitoba Teachers' Society (MTS) committees. I was also a featured teacher in two of MTS's television commercials, the first in 2009 and the second in the spring of 2019. I have also been lucky enough to be nationally recognized by the Indspire Awards in 2017, where I was awarded for being the Outstanding Education Role Model.

I am extremely lucky. I am lucky in all aspects of my life, and I certainly have nothing to complain about. What does feel contradictory, though, is talking about myself. In Anishinaabe culture, there are teachings and natural laws that we are to follow. Within the Seven Teachings we are taught to be humble and show humility. In writing about the aspects of my life, as rewarding as it is, it feels bizarre and unusual, almost unnatural. I have had these feelings throughout this work, but I consistently think about my end goal: to conduct this work with respect and courage for my children. I ask forgiveness and leniency from my ancestors and family, and I wish for them to see that I know what shoulders and shadows that I stand on.

My daughter Grace was born in August of 2011. Children are medicine. I sit on a Government of Manitoba committee that oversees the development of high school Indigenous studies curricula, and one of the members is an elder from Thompson, Manitoba. In explaining the life cycle of children into adults, he starts with the infancy stage of life and says, "Children are healers and they bring medicine, because when they are born, they immediately make us [parents] be better people."[22] Life took on new meaning the day Grace was born, and, like many things that have happened in my life, I was provided with more purpose. Grace is an old soul who is shy and loves playing. She is not particularly passionate about one thing, but she loves to sing, dance, visit family, be outside, snack, and play with her cousins. What more could you ask from your child?

My youngest daughter, Natalie, was born in May of 2015, and she is a fireball. She was brought into our world to provide us balance. She keeps my wife and me honest and challenges us every day in a multitude of ways. She is the jokester of the family and is always attempting to get a smile, laugh, or reaction from people in the room. She is quick to learn and observe human behaviour, and she can often be heard singing songs that are inappropriate, yelling expletives, crying irrationally, and stubbornly testing our patience. With this in mind, she is also very shy, but once you get a feel for her energy, and she recognizes that you notice her, she reveals her sense of humour and wit.

These young ladies have changed my world and are now my purpose. Obviously, my wife and I, and our schedules, revolve around our children's schedules. They keep us busy, as is the case for many other families, but my children bring me pure joy. As challenging as they can be as they grow and learn about our world, I feel like it is my responsibility to ensure they understand themselves, who they are, where they come from, so that they are able and prepared to answer where they are going and why they are here. As I continue to work for my community, always keeping their best interests in mind, it is Grace and Natalie that provide reason. They are the reason I do this work. This is *who I am.*

Home

You know, not so long ago, there were lots of people that did not need much to be happy. They were loved, and felt loved every day. That is what every person needs. There was a young girl named Margaret who loved to be loved. She used to live on the shores of Centre Lake, which is near the small Métis community of Vogar.

When she was small she would spend time with her mom walking along the beach at Centre Lake. Margaret thought it was the most beautiful place in the world. She would hold her mom's hand and simply walk in silence. She would pick rocks and show her mom; she would point at birds; and sometimes, she would just watch her mom's feet.

Afterwards, they would go back to their small but warm one-room house. In this house was a fireplace, a kitchen, and beds. The small house was built by her dad. There is no better feeling than going home.

NOTES

1 Chelsea Vowel, *Indigenous Writes: A Guide to First Nations, Métis, and Inuit Issues in Canada* (Winnipeg: HighWater Press, 2016).
2 *Truth and Reconciliation Commission of Canada: Calls to Action* (Winnipeg: Truth and Reconciliation Commission of Canada, 2015), 1.
3 Murray Sinclair, "Speech," Mosaic Institute Peace Patron Awards Gala, Toronto, ON, 18 May 2016, https://mosaicinstitute.ca/videos/.
4 Sinclair, "Speech."
5 Sherman Alexie, *The Absolutely True Diary of a Part-Time Indian* (New York: Little, Brown and Company, 2007), 166.
6 See *An Act to Encourage the Gradual Civilization of Indian Tribes in this Province, and to Amend the Laws Relating to Indians*, 3rd Sess., 5th Parl. (1857), https://caid.ca/GraCivAct1857.pdf
7 Thomas King, *The Truth about Stories: A Native Narrative* (Toronto: House of Anansi Press, 2003); Vowel, *Indigenous Writes.*
8 Chimamanda Ngozi Adichie, "The Danger of a Single Story," TEDGlobal, July 2009, https://www.ted.com/talks/chimamanda_adichie_the _danger_of_a_single_story?language=en.
9 Linda Tuhiwai Smith, *Decolonizing Methodologies: Research and Indigenous Peoples* (London: Zed Books, 1999).
10 King, *The Truth about Stories*, 128.
11 Vowel, *Indigenous Writes.*
12 Maria Campbell, *Halfbreed* (Toronto: McClelland and Stewart, 1973); Paulette Regan, *Unsettling the Settler Within: Indian Residential Schools, Truth Telling, and Reconciliation in Canada* (Vancouver: UBC Press, 2010); Vowel, *Indigenous Writes.*
13 Robin DiAngelo, *White Fragility: Why It's So Hard for White People to Talk about Racism* (Boston: Beacon Press, 2018), 28.
14 DiAngelo, *White Fragility*, 11–14.
15 Regan, *Unsettling the Settler Within.*
16 Sinclair, "Speech."
17 Wab Kinew, *The Reason You Walk* (Toronto: Penguin Canada Books, 2015).
18 Kinew, *The Reason You Walk*, 268.
19 Sinclair, "Speech."
20 Campbell, *Halfbreed*, 13.
21 Desiree Dorion, "Like a Flower," track 8 on Desiree Dorion and Joel Schwartz, *Tough Street*, 2007, compact disc.
22 Personal communication, 24 September 2028.

PART TWO

RELATIONS

Lake One Trail: Exploring the Egheze Kue Aze (Egg Lake) Landscape in Wood Buffalo National Park of Canada

LAURA PETERSON

Introduction

Wood Buffalo National Park (WBNP) is recognized for its international contributions to nature, including the largest free-roaming bison herds in the world. Yet, the park landscape is criss-crossed by multiple overland Aboriginal trails used by those who have stewarded the lands for millennia. As a collaborative research project, the Lake One Trail study explores the relationships that some Cree and Chipewyan families have with one of these overland trails in WBNP. The trail leads to one of the most abundant food source areas, and links historic settlements, camps, resource-harvesting areas, and spiritual and culturally significant places along its diverse landscape.

Trails have many purposes – they help us get from one location to another, and they lead to sacred and necessary places. Travelling a trail entails more than a physical journey; it results in teachings, stories, and connections between people and the landscape. Using an interdisciplinary approach incorporating local knowledge, storytelling, mapping, air photos, archaeological methods, and archival research led to the documentation of significant cultural values, perspectives, and teachings shared along the trail. As a land-based methodology it demonstrated the important role that individual and collective knowledge can have as one moves through and experiences the landscape, providing greater respect and understanding of the Indigenous contributions to the conservation of the land, and the unique cultural landscape within the park.

I was born in Vancouver, British Columbia, with ancestry from the Stó:lō and the Tahltan nations and adopted into the K'ómoks First Nation on Vancouver Island at the age of two. Throughout my life I have been grateful for the opportunities to visit and work with elders and knowledge holders, both on Vancouver Island and in northern Alberta and

Figure 6.1. Laura Peterson and Lawrence Vermillion walking the Lake One Trail during field work.

Source: Courtesy of Donalee Deck, 2014.

the Northwest Territories (NWT). Whether having tea while visiting in the community or being out on the land, I am grateful for the stories and teachings that were shared. These memories and connections have helped ground me as I journey through my own life experiences to understand my roots and where I come from.

My career at Parks Canada began in 1998 when I moved to the NWT and found myself immersed in Dene, Cree, and Métis cultures and territories. Based in Fort Smith, the headquarters for Wood Buffalo National Park (WBNP), we began to conduct a series of community cultural workshops with each individual Aboriginal group[1] to better understand their unique history and connection to the lands in WBNP, including important place names and stories. One of the concerns raised by harvesters and park users was the deteriorating condition of traditional overland trails in the park. As shared by the late Joe Vermillion, "people used to cut across country instead of going by river" (interview, 23 August 2008). Overland trails provide access to ecologically and culturally rich harvesting areas and are even more important when there is limited road or boat access. Poor trail conditions further contribute to the long history of displacement of people from the park, and these barriers to travel in the park continue to have impacts on the livelihoods of the people and the community over time.

These overland trails are disappearing with little information recorded about them, and few have been mapped. To help us understand the threatened state of trails, Robert Moor's 2016 book *On Trails* explores the various types of trails in North America. Moor states that with European exploration came the widening of trail segments associated with Indigenous groups, "first to accommodate horses, then wagons, then automobiles." While much of this trail network is buried or overgrown, "remnants of the old trail system can still be found when you know where – and how – to look."[2] However, many routes are fading from living memory or have already physically disappeared; this is the result of multiple causes, but especially changing land uses, evolving travel technology, and decades of extreme wildfire due to climate change, to name a few. As a result, there are few individuals today who hold knowledge of these overland trails, and even fewer who have experience travelling on them. The people who once travelled these trails are getting older, and many no longer have the physical stamina to reopen the trails.[3] With fewer people travelling on the land, the younger generation is growing up without learning about these trails, their family traplines, or the knowledge associated with them. Economically, trapping, hunting, and fishing are no longer viable as a sole means to support your family, and most people must supplement land-based activities with other work. Many people have left their community to obtain employment, including jobs related to the oil sands operations to the south of the park. These are some of the concerns expressed during the workshops, and they result in less time spent on the land between elders and youth, which has traditionally been the context in which knowledge, skills, and experiences related to elders' unique histories tied to the trail networks are passed to younger generations.

The trail research described in this essay was exploratory, aiming to research and capture important knowledge of trails before it is lost. As a pilot project, the Lake One Trail helped to understand how we can study historic overland trails in WBNP and what can still be learned about them. By documenting one trail, I learned about the connection of places along the trail and the larger regional network of trails, which together contribute to our understanding of Aboriginal use and occupancy in the region. The Aboriginal cultural landscapes of WBNP can mostly be considered an "associative" cultural landscape,[4] and what Ingold defined as "dwelling" on the landscape. People's activities on the land over time give meaning to place and are evident in oral traditions, place names, and "ways of knowing" and being.[5] Through people's relationship with the land, knowledge is gained by the direct experience of travel, by establishing a pattern of activity on the land that is seasonal in nature, and by engaging in a wide range of activities on the land.

This essay summarizes the findings of the Lake One Trail study and profiles the unique connections that three families have had with the trail and surrounding area over generations. This is preceded by a brief introduction to WBNP and the study area, followed by a description of the Lake One Trail and its seasonal relationship to the families connected to it through living memory and archival sources. This leads to a discussion about the concept of Aboriginal cultural landscapes and how trails, as one aspect, really go beyond their functional purpose of getting from one place to another. As a pilot project, Lake One Trail can be described as part of a living or evolving landscape, and thus helps us to understand people's connection to the trail over time. While conserving stories, teachings, and relationship to place for future generations, the study also demonstrates people's contribution to the conservation of this land.

Wood Buffalo National Park

Canada's largest national park, exceeding the size of Vancouver Island, WBNP straddles the Alberta-NWT border. It was first established in 1922 to protect the last remaining herds of free-roaming bison in Canada. Additional land in the south was annexed in 1926,[6] bringing the total area to 44,800 km². The park accounts for approximately 20 per cent of the current land mass of Canada's national parks.

In 1982, the park's two largest wetlands, the Peace-Athabasca Delta and the Whooping Crane Nesting Area (see figure 6.2), were declared Wetlands of International Importance under the Ramsar Convention on Wetlands.[7] This was followed by its designation as a UNESCO World Heritage Site in 1983, for both biological and geological reasons:

> [WBNP] … includes one of the largest free-roaming, self-regulating bison herds in the world, the only remaining nesting ground of the endangered whooping crane, the biologically rich Peace-Athabasca Delta, extensive salt

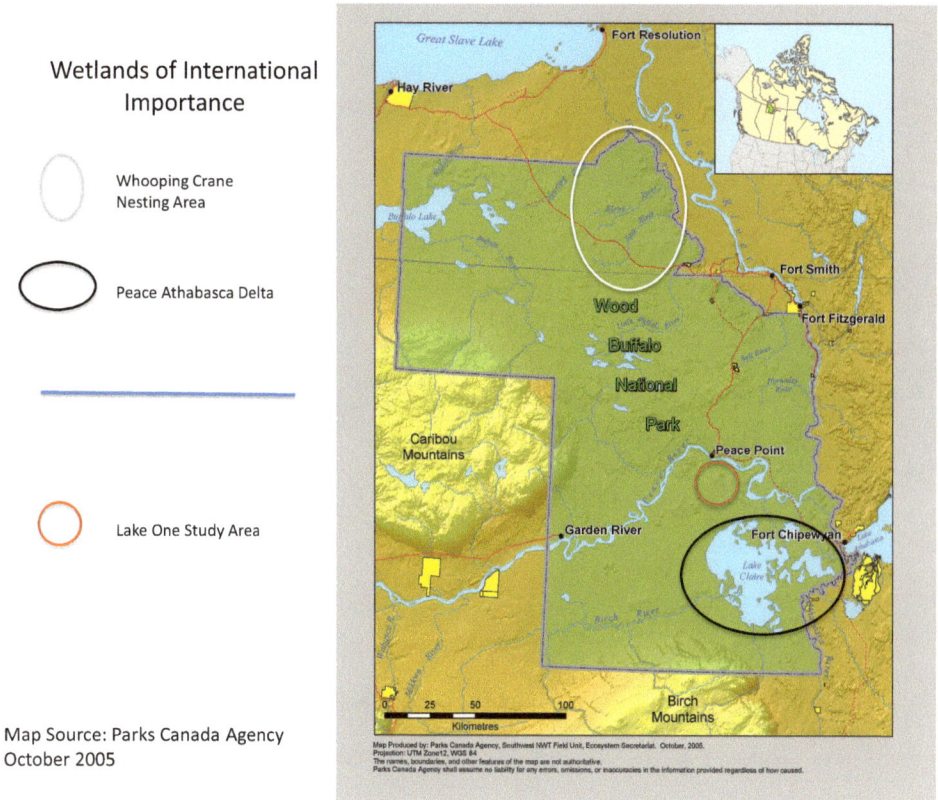

Wetlands of International Importance

Whooping Crane Nesting Area

Peace Athabasca Delta

Lake One Study Area

Map Source: Parks Canada Agency
October 2005

Figure 6.2. Approximate locations of Wetlands of International Importance in Wood Buffalo National Park in relation to the Lake One study area.

Source: Courtesy of Craig Brigley.

plains unique in Canada, and some of the finest examples of gypsum karst topography in North America.[8]

The park is a vast mosaic of boreal grasslands, wetlands, and forests, with numerous rivers, creeks, lakes, and ponds, and it currently stands as the only place where the predator-prey relationship between wolves and wood bison has continued, unbroken, over time.[9] In short, there is significant international recognition of the park's natural resources.[10] However, formal recognition that respects and recognizes local community values and the long-standing relationships that many nations have with the park lands over time has yet to be acknowledged in this way.

Figure 6.3. Winter trail in Wood Buffalo National Park.

Source: Conibear Lake Trail 20-Nov-32, UAA-1979-021-035-418.

WBNP differs from most of the other early national parks in that it allowed, albeit reluctantly, some Aboriginal and other occupation and use to continue after it was created. Nonetheless, the intimate knowledge that people held of the land and its resources was mostly ignored during the establishment and early management of the park, even though local knowledge of trails facilitated botanical and faunal investigations in its first years.[11] Because of the park's size and limited road access, both summer and winter trails were key to accessing its interior and other hard-to-reach areas.

Prior to the park's establishment, many generations maintained overland trails, and knowledge of these trails was accumulated through their experiences and stories passed down from one generation to the next. As a result, many of the pack trails – for horses, cat trails,[12] wagon roads, and warden trails – that were identified in the park were initially traditional trails. As well, the sharing of local knowledge between generations kept these trails open by continued use over time. Today, loss of this

Figure 6.4. Summer-fall trail in Wood Buffalo National Park.

Source: Bison Trail in Wood Buffalo Park, 9-Sept-33 UAA-1979-021-035-566.

knowledge threatens this long-standing Aboriginal relationship with the park. An early document written in 1923 provided observations on the newly formed park. It gave a description of the buffalo, their habitat, range, and threats, along with the flora and fauna of the park. Of particular interest was the reference to buffalo trails, including their extensive network in the park and their use as pack trails for warden activities.[13]

The study area is known as the Lake One Prairie and contains a series of lakes that were at one time a single, large body of water[14] (see figure 6.5). In the Dene language, Lake One is known as Egheze Kue Aze, or Egg Lake, with "Aze" meaning small or little.[15] In addition to providing drinking water, this lake attracted nesting migratory waterfowl, so people travelled there to collect eggs. The open prairie was also important habitat for bison and other ungulates. The families who traditionally used this area indicated that a trail referred to as "Lake One Trail" provided access to this important water source located between the Peace River and Lake Claire, two additional important travel and water sources.

Orientated in a north–south direction, the Lake One Trail may be one of the oldest known trails in WBNP. Situated between two known archaeological sites, Peace Point and the Lake One Dune site, the trail has been subject to long-term use by different groups. Previous archaeological investigations and traditional knowledge have indicated this area has been used for millennia.[16] This trail study focused on just over a century of land use and travel on the trail, including living memory between 1920 and 2015, and with knowledge of trail use dating back to at least 1899.

While the exact age of the Lake One Trail is unknown, it was remembered as being used by Charlie Simpson's grandparents' generation, which takes the trail back in time to the late nineteenth century, and possibly much earlier.[17] During the time of the informants' parents and grandparents, about twenty families would camp around Lake One in the springtime. "People camp any place – usually the higher poplar ridges" (Fred Vermillion, personal communication, 12 May 2014). Fred also talked about ancient campsites located on the sand hill along the south shore of Square Lake. "People just camped all over ... they were mostly Chip [Chipewyan], but also Cree. There was lots of muskrat and lots of trappers long ago. My dad and uncle trapped here" (Lawrence Vermillion, interview, 7 October 2014). Today, the Lake One Prairie is part of Group Trapping Area 1209,[18] used primarily by three local families: the Cheezie, Vermillion, and Simpson families. I worked closely with individuals from these families as they shared their experiences travelling on the trail and their family connections to the study area.

The Lake One Trail

Lawrence and Fred Vermillion, Lawrence and Philip Cheezie, and Charlie Simpson all described the Lake One Trail as an overland trail that intersected many other trails.

"Many trails have started as walking trails, but they evolved into dog team trails, then skidoo and even cat trails," explained long-time former park employee Gordie Masson (interview, 23 June 2014). The Lake One Trail indeed was an old walking and later dog team trail that had been turned into a cat trail. Local parks employees widened the trail with a cat in 1965, and the original route was chosen using local knowledge and experience. The trail was used by trappers and later for park operations. The Lake One Trail branches off into more than one dog team or skidoo trail, and there is more than one access point into the Lake One Prairie, depending on which direction the traveller is coming from.

For this trail study, we documented one route of the Lake One Trail, approximately eight kilometres in length, that starts with the "cat road" at the Peace River and ends with a "cat road" at Wolf Ridge.[19] Cat roads are usually straight, created with machines that allowed other types of snow vehicles or trucks to travel on the trail and carry supplies and

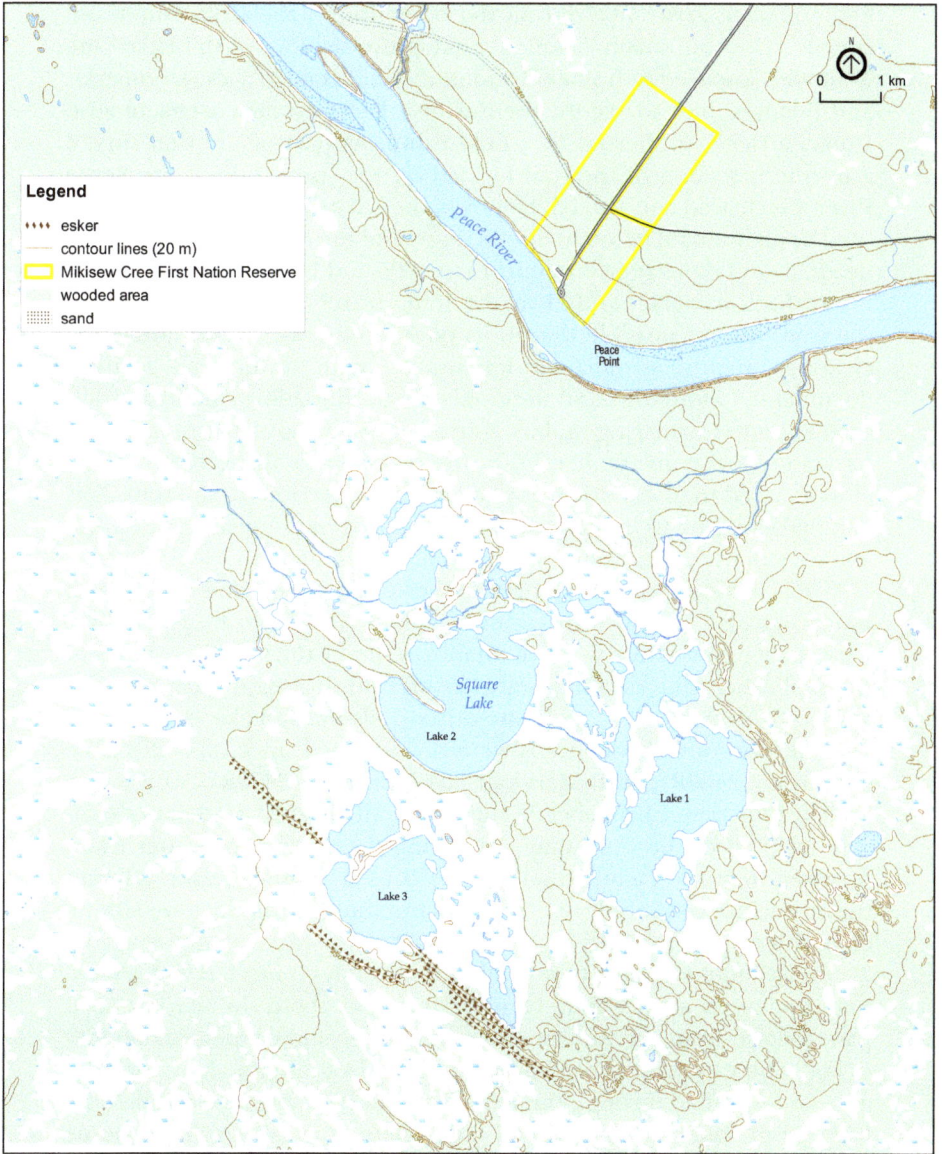

Figure 6.5. Lake One Prairie study area.

Source: Parks Canada.

equipment to areas, like during the operation of the Lake One bison corral.[20] The trail survey entailed walking two sections of the Lake One Trail with knowledge holders to identify important places, landmarks, and features, and to record them with a Parks Canada archaeologist, Donalee Deck, supported by a field camp. As part of the trail survey, a northern section of the trail located on the south side of the Peace River was cleared and walked. However, the Lake One Trail had not been used or maintained in ten to fifteen years, and the overgrown vegetation, deadfall, and damage from wildfires in 2005 and 2007 made it challenging to find, navigate, and clear the trail. Travelling the full length of the trail was not possible due to its poor condition, as even after trail clearing, much of it remained impassable. Some of the main features documented along the trail were old cabin foundations, ancillary trails such as those for snaring rabbits, spiritual sites, campsites, and resource-harvesting locations (table 6.1). In the process of walking the trail, the elders shared their experiences and family connections to the trail. This entailed stopping to build a fire, make tea, and have lunch while stories and experiences of travelling on the trail were shared.

A trail section that ran along an esker at the south end called Wolf Ridge was also walked with family members, thanks to transport by a Parks helicopter. While we were unable to survey the middle sections of the trail, where the cat road branches off into dog team trails, we were able to fly over this area and note other trails in the Lake One Prairie. Overall, portions of four trails were documented, including the Lake One Trail (2469R), 30th Meridian Baseline Trail (2470R), Wolf Ridge Trail (2490R), and Lawrence Vermillion Trail (2515R) (see figure 6.6). The majority of the trails mapped during the study were winter trails; however, they also included walking trails used at other seasons. Today, these trails are commonly referred to as traplines, as traplines are all situated with respect to trails.

Stories about the Lake One Trail were strongly connected with a spiritual site known as the Little Man. Information about the site has been passed down from previous generations by Cree and Chipewyan families in Fort Chipewyan and Fort Smith, especially by those who have used the area and travelled through the Lake One Prairie. Lawrence recalled the Chipewyan name for the spiritual site as Deneza, meaning "little people living in hollow trees" (Lawrence Cheezie, personal communication, 26 January 2014). There are numerous stories about the "little people" who provided protection to this place, symbolized physically on the landscape by a carving on a tree stump. At this place, the traveller would stop and visit and show respect by leaving offerings for a safe journey through the country. "The visitor asks for assistance or offers gratitude

Table 6.1. Summary of heritage features at Egheze Kue Aze (Egg Lake)

Heritage features in the study area	Quantity
Precontact Sites	**2**
Lake One Dune (34R1)	1
Peace Point (30R1)	1
Cabins	**8**
John Simpson cabin (34R197)	1
Lake One cabin remains	6
Isidore Simpson cabin (34R197)	1
Cabin foundations	**4**
Simpson and Vermillion cabin foundations (34R197)	3
Peace Point cabin, also known as Cabin #2	1
Historic Settlements	**2**
Peace Point settlement	1
Jackfish settlement	1
Historic tent camps	**2**
Parks tent camp (2467R) and traditional tent camps	2
Camp associated with families who trapped at Lake One long ago	Unknown
Lunch or picnic spot	**1**
Rest stop on the way to Fort Chipewyan by dog team	1
Semi-subterranean dwelling	**2**
Pit house or dwelling	2
Spiritual site	**4**
The Little Man (2338R)	1
Child's burial	1
Anticline; "the church"	1
Red Rock Island; Drum Island; Red Stone Island; Firestone Island	1
Total	**25**

Source: Knowledge Holders and Field Survey

for successful trapping or hunting by paying respect to the Little Man and leaving an offering of value. In return, the Little Man looks after this place, to ensure there is clean and high water levels and abundant muskrats" (Lawrence Vermillion, interview, 11 October 2015). The Little Man can provide guidance to the people who travel the area, but it can also do harm if not respected or acknowledged appropriately. In addition to the offering site, there are other landmarks in the study area connected to the story of the "little people" where stories and teachings are shared.

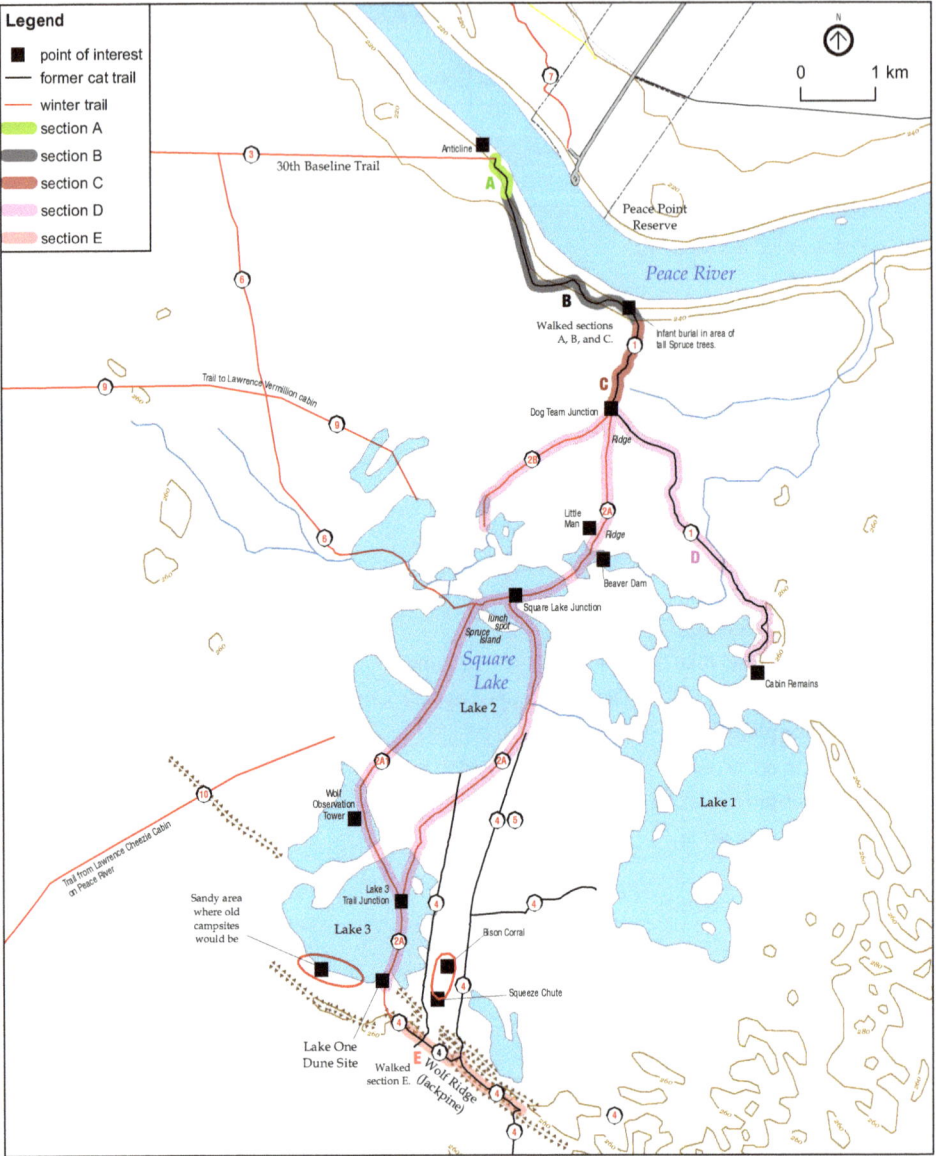

Figure 6.6. Lake One Trail in the study area. Lake One Trail sections (north A–C; middle D; south E): (1) Lake One cat trail; (2A) Lake One winter trail; (2A1) Lake One winter loop; (2B) Dog team trail to lakes north of Square Lake; (3) 30th Meridian Baseline; (4) 1948 cat road; (5) section of L. Vermillion lynx trail that follows cat road; (6) L. Vermillion road; (7) 1958 winter road; (9) trail to Lawrence Vermillion cabin; (10) trail to Lawrence Cheezie cabin on the Peace River. Note that trail 8 is not situated in the study area.

Source: Courtesy of Craig Brigley.

Lake One Trail Seasonality

The trail has been used year-round to access a variety of resources, but more often during the spring to hunt muskrats, beavers, ducks, and collect goose eggs. Other important resources harvested included bison (before the park's establishment), barren-ground caribou, moose, black bear, deer, grouse (known as chickens), migratory waterfowl (ducks, geese), fish, and an abundance of other fur-bearing animals that included wolves, wolverine, lynx, fishers, marten, mink, rabbit, and squirrel. The most abundant resources harvested in the Lake One study area were fur-bearing animals (figure 6.7). In addition, both high and low bush cranberries, raspberries, saskatoons, and blueberries were known to have been foraged along the trail, including medicinal plants.

Lawrence Vermillion (interview, 7 October 2014) explained, "During the winter we trap fine furs such as wolf, lynx, marten, fishers, minks, squirrels, and weasels. Fine furs are also known as long-haired fur." Occasionally his dad would travel alone to Fort Chipewyan, especially during specific times of the year: Christmas, the end of February, Easter, and in May. They sold furs and bought a few goods. They sold the highest number of their winter furs at Christmas time, and at the end of February after trapping season closed for fine fur.

Barren-ground caribou and white-tailed deer were in the area only in the winter months. The caribou came in November after freeze-up, moving into the forest to feed on lichens and take shelter from the wind. They migrate annually between forest and tundra. Along the 30th Meridian Baseline Trail (see figure 6.6) the hills were covered with lichen, good winter habitat for barren-ground caribou. "When the caribou came we [would] shoot them right from our cabin [on the Peace River]; my mother [Madeleine, née Tourangeau] was a good shot with a .30-30" (Lawrence Vermillion, interview, 2 February 2015). With the variety of resources at all times of the year, I can understand why there are stories of the trail being accessed by anyone in need of food at times when resources were sparse (Huisman, interview, 29 May 2014).

Family Connections to Lake One Trail

Brief family profiles provide context for the knowledge that each elder shared, which derives from their individual experiences. Knowledge about the trail and the area is not necessarily a collective knowledge, but one that is specific to each individual's experience and his/her family's long-term connection to this place. For example, a consistent theme throughout the generations was use of the Lake One Trail during particular seasons for harvesting.

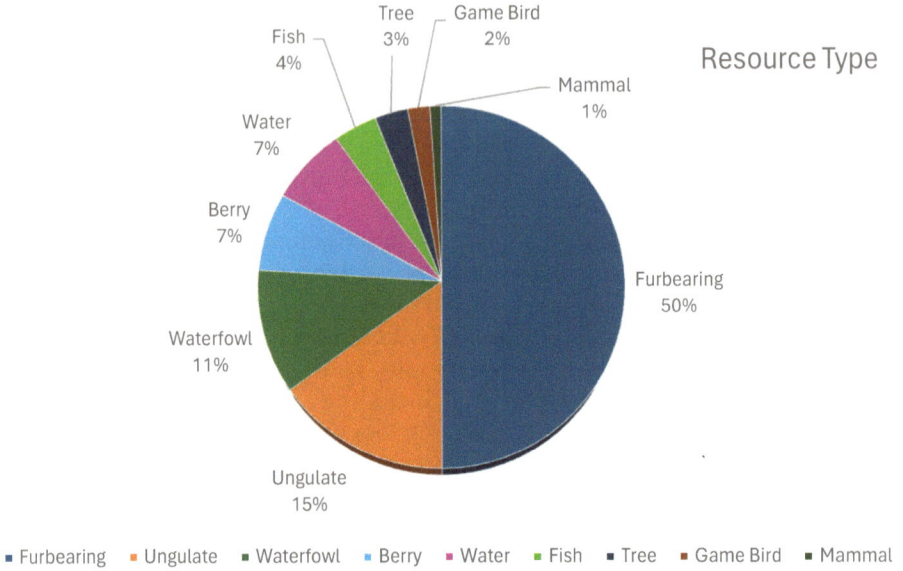

Figure 6.7. Resources harvested within the study area.

Source: Courtesy of Laura Peterson.

Family History of Lawrence and Fred Vermillion

Brothers Lawrence and Fred Vermillion grew up with twelve other siblings. Joe Vermillion, one of their brothers who also travelled through this area, passed away in 2014. Their father was Alexandre Vermillion from Jackfish River area, who married Madeleine Tourangeau, a Métis from Fort Chipewyan, in 1936. Her parents were Pierre Tourangeau and Mary Rose (née Mercredi). Pierre was Métis from Red River, Manitoba, and spoke mainly French but also some Cree. Pierre passed away in 1958.

The Vermillion family was originally from the House Lake Settlement located in WBNP.[21] Their paternal grandmother, Bienvenue Vermillion (née Cayen), and her children moved to the Jackfish River Chipewyan community on the Peace River due to deaths in the settlement in the 1920s. Alexandre Vermillion was very young at the time. The family lived a traditional nomadic way of life, moving in an area from Poplar Island on Lake Mamawi to the Coupé River and Fort Chipewyan, before settling on the Peace River. Lawrence Vermillion had the opportunity to spend time with his paternal grandmother, Bienvenue, on the Peace River and various places around Fort Chipewyan before she died in 1960. She enjoyed living off the land in the bush, snaring rabbits, shooting grouse, setting fish nets, and trapping squirrels in the winter. Lawrence described in detail how she made the snare a particular way and how it killed the rabbit quickly, thereby ensuring the meat tasted better. In answer to my question about how his grandmother learned these skills, he replied, "she must have been raised that way as daughters were taught how to set snares and catch squirrels in the winter." Lawrence also raised his daughters this way. Bienvenue lived in a tent most of the time, near where Lawrence's family lived.

When walking the trail with family members, we learned that Lawrence's grandmother had a cabin near the Lake One trailhead (Bienvenue Vermillion cabin foundation, A3c in figure 6.9). The remains of her cabin are comprised of a cellar depression and some family items that Lawrence recognized. He pointed out his grandmother's water pail and brother Fred's chainsaw blade hanging in a tree north of the cabin foundation. There were also axe cut blazes on three trails south-east of the foundation marking where she set rabbit snares, which would have been used all year round.

The cabin was twelve by fourteen feet and built in 1953. Lawrence shared that the cabin was here only two years before it was dismantled and floated downriver to Baril Creek, where Bienvenue's grandson, Albert Gladue, was living. Lawrence explained that an Indian agent at the time built the cabin with the help of Isidore Simpson, but that they had done so in the wrong location. Bienvenue wanted to be closer to her family at the mouth of Jackfish River, where she had a tent, and she only spent a few months in the newly built cabin.

Lawrence Vermillion

Lawrence Vermillion is a member of the Mikisew Cree First Nation in Fort Chipewyan, Alberta. He was born 28 February 1939, delivered by a midwife on the Peace River, just below the Jackfish River in WBNP. This area was a traditional spring and summer fishing and gathering place that was also known for the abundance of large game in the area. Lawrence attended the Holy Angels Catholic School in Fort Chipewyan when he was seven years old and lived in the convent (i.e., the residential school). After attending school for seven years, at the age of thirteen he returned to his family on 30 June 1952.

Lawrence said that his father, Alexandre, used a twelve-foot spruce-framed canoe, also known as a "rat canoe," for hunting in the lakes and sloughs for muskrat. Alexandre walked or carried the canoe one way, from the Peace River to Lake One, to hunt ducks and collect eggs. Then he waited until the spring thaw before being able to paddle on the lake when the water levels were at their highest. The spring muskrat trapping season was described as lasting anywhere from two weeks to a month, with the season starting as early as March. "In March the muskrats are fully grown and fat with thicker hides, thicker fur, and sold for a better price. Besides trapping we would also shoot muskrats with a .22 rifle. Trapping the muskrat meant that they drowned right away and were better eating and good for dry meat."

After open water, Lawrence's family travelled to Lake Claire by boat to "catch the thousands of birds in the bays." They hunted ducks, geese, and wavies (i.e., snow geese). Once at Lake Claire, trappers from all over, but particularly from Lake Claire, Hay River (also known as Prairie River), and the Peace River areas, spent late spring and summer in Fort Chipewyan, at the Quatre Fourche and Lake Mamawi fishing whitefish and making dry fish. This was also a time when Lawrence's dad would travel to Edmonton for a month or more in June or July to visit friends. He travelled by boat to Fort McMurray and then by train. The first time Lawrence travelled on the Lake One Trail was with his family in 1953.

The 30th Meridian Baseline Trail (2470R) is about fifty meters behind and to the west of the Simpson homestead (A4, figure 6.9). The access point is still visible, though it is overgrown with vegetation. Lawrence guided us to this trailhead, and he called the trail his "road," which also meant his trapline. This trail leads to his cabin just above the Boyer Rapids on the Peace River. He also referred to the trail as a "portage" between his cabin and the Simpson homestead. It is the "main road" from Lawrence's cabin and runs straight "across country." Philip Cheezie (interview, 27 February 2014) remembered this trail as "a nice wide trail, but the trail is gone now, it burnt. Before the fire, you could travel all the way to Fort Chipewyan in the wintertime from my cabin on the Peace River [using a portion of the 30th Meridian Baseline Trail]." Government surveyors cleared the baseline when Lawrence Vermillion's father was seventeen years old (ca. 1919), just after the First World War had ended. During the field survey we located a long brass pin with a flat circular top in the middle of the trail, and around the pin were four deep holes marking its location. As described by Lawrence Vermillion (interview, 8 October 2014),

> They surveyed a brass pin every mile using six horses packing their supplies and equipment and cleared the baseline 10–12 feet wide. The [original] purpose of the trail is still unknown, but trapping was good some years on

this trail for lynx, marten, and fishers. It was also a good area to find cranberries and moose berries [also known as highbush cranberries].

Lawrence was the last person to clear this trail, twenty years ago. At the time he also cut a "road" from the 30th Meridian Baseline Trail directly across to Lake One. Lawrence Vermillion recalled using dogs on the trapline until about 1980, when snowmobiles became more common. He described one of the first snowmobiles to come out in the 1970s as a single cylinder trapper's model that was very slow. Then in the 1980s a smaller and lighter model came into use, which he described as "fabulous." Soon, the snowmobile became more popular and eventually essential for winter bush travel. As a result, dog teams became more of an icon and connection to the "old ways."

In 1978, Lawrence went back to full-time trapping and lived in the Jackfish River area (Group Trapping Area 1209) with his family. They lived there year-round and continued trapping and hunting for a living until 1986. Lawrence continues to spend much of his time on the land at his cabin along the Peace River, travelling in the park often with his grandson and other family members.

Fred Vermillion (1940–2021)

Fred Vermillion was a member of the Mikisew Cree First Nation in Fort Chipewyan, Alberta. He was born 8 March 1940 in Fort Chipewyan, with the help of Jenny Flett, who was a well-known and respected midwife. He grew up with twelve brothers and sisters, but one older brother was lost in the bush at two years of age, and an older sister passed away when she was a baby.

Fred attended the Holy Angels Catholic School in Fort Chipewyan when he was five years old and lived in the convent for ten months of the year. During the summer months he was back with his family and travelled by boat out on the land. His grandparents travelled with them, and Grandmother Bienvenue taught him how to snare rabbits and skin a moose and other bush skills. He mentioned that by this time she was hard of hearing, which made it difficult to communicate with her. They would travel all over Lake Mamawi, Lake Claire, and up the Peace River living on wild meat, bear, and moose.

After attending school for eight years, he reached grade 7 and returned to his family full time in 1953. At this time, Fred stayed home for a couple of years and trapped with his mom and dad in the bush until he got his first summer job at the age of fifteen. Fred returned to Fort Chipewyan in the spring of 1972, trapping in the winter at Jackfish River and working in the summer months for WBNP. He travelled long distances by foot and dog team in the park as a trapper and park employee.

The first time Fred travelled on the Lake One Trail was also with his family, in 1953, the same year he returned from the convent. He travelled with his family by dog team, leaving their cabin on the Peace River, travelling along the 30th Meridian Baseline Trail to Peace Point, up Lake One Trail to Lake Claire, Hay River (known locally as Prairie River), and then on to Fort Chipewyan, arriving home for Christmas. He made numerous trips to Lake One after returning to the Peace River area as an adult. Fred enjoyed spending time on the land at his cabin on the Peace River with his family.

Family History of Charlie Simpson (1948–2020)

Charles Willard Simpson was a member of the Mikisew Cree First Nation in Fort Chipewyan, Alberta. He was born 28 November 1948 in Fort Chipewyan, delivered by a midwife. Charlie's mother was Cecelia Cardinal, a Métis from Fort Chipewyan. Her father's family had followed the fur trade to this area from Lac La Biche, Alberta. Her parents were David and Madeline Cardinal. Charlie's father, Archie Simpson, was born in 1917, and was a well-known Chipewyan trapper and hunter. He married Cecelia in 1944.

Charlie's paternal grandfather, Isidore Simpson, was a member of the Fort McKay First Nation until he and his family moved to the Peace River area and transferred to the Chipewyan Band to continue trapping in the park.[22] He married Corone Benoit, a Chipewyan from Fond-du-Lac, Saskatchewan, and they had ten children. In the family tradition, Archie would say to Charlie that your "Mushum" (grandfather) was a good "hunter" and "provider."

Several features associated with the Lake One trailhead are collectively referred to as the "Simpson homestead" (A2, A3a–c, A4, and A5 in figure 6.9). Isidore Simpson's one-and-a-half-storey cabin is the most evident and well known. Adam Dene and Isidore Simpson built it in 1933, using axes and dovetailed construction[23] (with the help of Isidore's sons, who lived nearby). They replaced an older cabin that was destroyed by a flood in 1932, and the newer building sits directly on the older foundation. Adam Dene was a Chipewyan who lived just up the Peace River beside the Cheezie cabin, on the north side of the river. He was considered very skilled with an axe (these skills were passed down from his father and grandfather), and the cabin has been described as "well built by a local carpenter who took the time to do a good job".[24] Known locally as the Simpson cabin, it is one of only two historic cabins from this time period still standing today in the park.[25]

Figure 6.8. Isidore Simpson log cabin at the trailhead, October 2014.

Source: Courtesy of Laura Peterson.

Isidore Simpson and his wife Corone (née Benoit) had eleven children – eight boys and three girls – all of whom grew up in this house. Later, Charlie and his brothers and sisters all slept on the upper level, and he spent time here until he was six years old (1948–55). In 1955 his family moved out, and he "never returned to this house as a child." Isidore Simpson's son (and Charlie's uncle) Billy and his wife, Pauline (née Aze), and their large family were the last to live in the Simpson cabin.

Charlie was the oldest of five children raised at Peace Point, though at the age of six, Charlie attended residential school at the Holy Angels Catholic Church in Fort Chipewyan. He was there for ten years, during which time he lost his ability to speak the Cree language. While at the residential school he was home only during the summer months, and when he did have an opportunity to go out on the land, it was spent on his dad's trapline, known as "Archie's Road,"[27] originally blazed by Charlie's grandparents in the 1920s. Before travelling to their trapline for the winter, Charlie remembered how his family prepared food for their dogs and for themselves in the fall. They hung fish to freeze, they snared

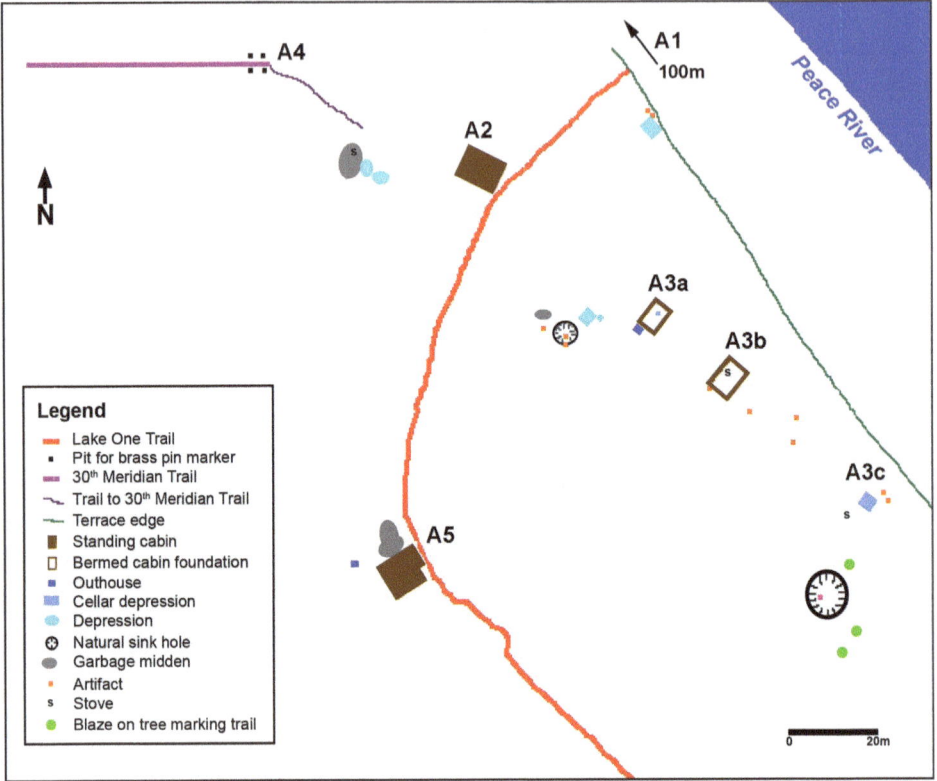

Figure 6.9. Historic features at Lake One trailhead. Simpson's cabins site (34R197) along the Peace River and Lake One Trail (2469R): (A1) karst anticline;[29] (A2) Isidore Simpson standing cabin; (A3a) Archie Simpson cabin foundation; (A3b) Billy and Pauline Simpson cabin foundation; (A3c) Bienvenue Vermillion cabin foundation; (A4) 30th Meridian Baseline trailhead; and (A5) John Simpson collapsed cabin.

Source: Courtesy of Donalee Deck, 2014.

rabbits, and they hunted moose. "We trapped all fall for squirrel [to sell] for Christmas time. That's how young guys got started, with squirrels" (Charlie Simpson, interview, 16 May 2014).

Charlie's first time on the Lake One Trail was in the spring of 1968, when he was nineteen years old. He did a spring muskrat hunt with his dad at Lake One for a month in May. Charlie remembered trapping when the ground was still frozen and portaging a canoe. They walked the trail in the early spring, walking behind dogs pulling a sleigh. Some dogs carried packs. During the spring hunt they stayed in a tent. Following the cat trail for three to four kilometres led them to the eastern edge of the Lake One Prairie and the remains of two older cabins that date back to before his dad's time. Semi-subterranean dwellings were also supposed to be in the vicinity. None of these features were located during the trail survey. However, Charlie remembered them well, because the two cabins were located halfway to Lake One and were used as a navigation marker. Charlie also mentioned how they would pass his cousin's gravesite when they walked along the trail to Lake One. His cousin was buried in the late 1940s or early 1950s as a very small child; Charlie was not sure how he died, but he stated that at the time people were just buried where they died, out on the land. We were unable to locate the gravesite due to the thick understory of fallen tree debris from the forest fires.

From 1978 to 1981, Charlie and his wife, Margaret Gladue, lived at Peace Point while their three children attended school there. At that time there were approximately ten families living at Peace Point. Charlie had four to five uncles who trapped on the same trapline at one time. One of those uncles, John Simpson, used to collect eggs at Lake One in the spring. The last time Charlie travelled on the trail to Lake One was in 1999 with his uncle. John worked for WBNP in various capacities and was the last one to live at the Simpson homestead in the 1980s. He travelled the Lake One Trail often but died in the winter of 2003 when he had a surprise encounter with a buffalo while skidooing on the Lake One Trail.

Family History of Lawrence and Philip Cheezie

Lawrence and Philip Cheezie's mother was Helene Aze, a Dene from Fond-du-Lac and Black Lake, Saskatchewan. Aze means "little," so she was known as "Little Helen." Her parents descended from the "Caribou Eaters," Chipewyans who continued to follow the barren-ground caribou. [28] Their father was Louison Tchize, a Chipewyan from Garden River and Fort Vermillion whose parents were originally from the House Lake settlement at Birch River in WBNP. Louison's father was Michel Tchize, whose last name, when spelled Tzhi'ze, means "Little Lynx." Helene and Louison married in 1924 [29] and had thirteen children, though many of the older siblings died during influenza epidemics in the 1920s and 1930s, before the family moved to the Peace River in 1935. Alexandre Vermillion (father of Lawrence and Fred) was already living at Jackfish on the Peace River, and he and Louison were close friends. Louison operated a Hudson's Bay Company store on the Peace River, and people came through by dog team to buy groceries.

Lawrence Cheezie (1946–2019)

Lawrence was a Chipewyan Dene and a member of the Smith's Landing First Nation in Fort Fitzgerald, Alberta. He was born 28 November 1946, delivered by his grandmother in a teepee near the mouth of Jackfish River on the Peace River. He remembers being at a place with a waterfall shaped like a horseshoe. Lawrence recalled spending time with his grandparents, "being packed on my grandmother's back, fed caribou tongue and hung on a tree in a moss bag, while she collected firewood in a dress that hung to her ankles, wearing fancy moccasins that were pointed" (Lawrence Cheezie, personal communication, 26 January 2014).

Louison Cheezie spoke of the Lake One Trail to Lawrence and Philip. While living at a settlement at the mouth of the Jackfish River, he would cut across the land to Peace Point by foot. On at least one occasion, in the summer, he walked all day along the 30th Meridian Baseline Trail from Jackfish to Lake One carrying his twelve-foot birch canoe and with his dogs carrying packs. The weather determined whether he would stay overnight at Lake One and how long he hunted or trapped.

Lawrence Cheezie attended the Holy Angels Catholic residential school for eight years until the age of twelve, and then moved to Fort Smith, NWT, in 1958. The first time Lawrence travelled the Lake One Trail was in 1980, with his dad and his brother Philip to hunt muskrats for two weeks in March, trapping rats along with the Simpson family. Lawrence explained, "the trail is well used because of the amount of wildlife in the area." When he walked the trail thirty to thirty-five years ago he could see a depression. "A depression in the ground showed how well used it was." Since then, the ground has been packed down and the depression has become less visible (Lawrence Cheezie, personal communication, 26 January 2014).

The Simpsons and Cheezies would visit back and forth with one another while they were trapping muskrat and beaver during the spring hunt. "We would leave our cabin and go to the spring camp for muskrat and beaver. You could walk between the two [locations] in a couple of hours. Lots of muskrat in late spring [mid-April] to the end of May. The whole family would be at the tent camp, where they slept on spruce boughs in a tent. They also camped on the trail and would usually go back to the same site, usually close to water or a drinking source on the lake" (Lawrence Cheezie, personal communication, 26 January 2014).

Lawrence also told a story about how his father once left him in the sled on the 30th Meridian Baseline Trail while he went looking for a lynx trap. Apparently, a caribou had stepped in their lynx trap, and his dad found the trap in the bush. The trail was described as having lots of lichen – a rich winter food source for caribou – before the trail was burned by forest fires and much of the lichen was destroyed. Lawrence quit trapping in 1983 because the fur prices were so low he could not make a living at it.

Philip Dolphus Cheezie (1932–2014)

Philip was Chipewyan Dene and a member of the Smith's Landing First Nation in Fort Fitzgerald, Alberta. He was born 30 March 1932, delivered by a midwife at Johnson Lake on the Birch River in WBNP. From there his family moved to a settlement near the mouth of the Birch River (delta) to a place he called Chi La Kue, meaning "Point Lake" in Chipewyan, and also known as House Lake (Philip Cheezie, interview, 10 May 2010).[30] His family lived at House Lake until he was three. His grandfather, Michel Tchize, was the chief at the time and had a North West Company store near the mouth of the Birch River. While living at House Lake, three of Philip's older siblings passed away – two brothers and one sister – from flu and smallpox epidemics.

When relocating to Jackfish, Philip and his family followed a winter overland trail by dog team from Lake Claire to Peace River. Philip spent most of his childhood out on the land, but he also attended mission (residential) school for a short time in Fort Chipewyan. He likely would have died in the mission had his dad not taken him out to live in the bush on the Peace River with his parents, where he trapped until he was eighteen or nineteen years old. In his younger days, he and his dad, Louison, were known to have the fastest dog team around. "They had well-trained dogs and took care of them really well" (Lawrence Cheezie, personal communication, 1 April 2015).

Philip described the trail from Peace Point to Lake One as an "old, old trail"' He knows the trail from hunting at Lake One for rats (muskrats), wolves, lynx, and moose, and collecting eggs from waterfowl. He remembers travelling the trail by foot – portaging a canoe from their cabin on the Peace River to the trail to Lake One, then paddling on the lake. Later, Philip also trapped in the French Lake area and travelled to Fort Chipewyan by dog team at Christmas and Easter; he described this trip as "a long ways to [Fort] Chip – two days travelling by dog team from Peace River."

Discussion

The personal profiles shared by each knowledge holder helped me to understand their unique family and individual connections to the Lake One Trail over many generations. This collective and intergenerational knowledge of the Egheze Kue Aze landscape demonstrates the importance of this area to their families and collectively to their communities. The trail study highlighted the important contributions that individual experiences, stories, and teachings shared on the land make to knowledge creation and knowledge transfer.

It is well known that landscapes evolve, and that people's relationship with the landscape also change and evolve as well. Cultural landscapes have been described as a living landscape by Andrews and Buggey,[31] who write that "while cultural landscapes change so do the cultures that commemorate them." I found this to be true for the Lake One Trail study in that it helped me to understand how the family relationship to the trail and to the Lake One Prairie evolved over time. The trail itself is a medium from which new knowledge is created, based on the reasons for travel, seasonal use, mode of travel, and existing conditions. When the trail is used regularly, by default the traveller helps to maintain the trail just by using it. This has become especially important as a result of climate change. Due to current impacts like increased wildfires, the importance of accessing and documenting knowledge of these ancient trails cannot be underestimated. For example, the Lake One Trail survey demonstrated the connections that one trail can have to the broader landscape and region due in part to the seasonal use of the trail for harvesting of important resources by many different groups over many generations. Overland trails such as the Lake One Trail play a pivotal role because they transcend time along with physical space and boundaries, promote greater access, and help future generations understand how different cultural groups came together to share resources and work together.

Those of us who walked the trail learned much more than we could have sitting in Fort Smith or Fort Chipewyan drawing lines and isolated points on a map with knowledge holders or looking at air photos and archival maps. Walking the trail was more than simple "ground-truthing." During the process, numerous places of cultural and spiritual value were identified that together provided a much fuller understanding of the cultural connections to the trail over time and allowed us to make important connections between these places and to people's relationship to the broader Egheze Kue Aze landscape.

The stories and experiences of travelling the Lake One Trail are held in the hearts and minds of the family members who used it. Being mobile was a large part of a trapper's family history. The act of moving through the landscape and the interactions that occur in the process allow the traveller to create knowledge of the trail on which he or she is travelling. As new knowledge and traditions are passed down between generations, the trail itself, along with people's relationship to it, evolves and is being continually shaped by those who spend time on this landscape.[32] Similarly, the Lake One Trail survey helped me understand the importance of accessing the Lake One Prairie through the process of travelling on the trail itself.

Echoing the words of my late friend Charlie Simpson, there is a need to think "outside the box" to preserve his family's trails like Lake One and "Archie's Road." This action alone would make an important contribution to the way the largest park in Canada is managed, now and into the future, for his grandchildren and great-grandchildren. Parks Canada has the land base that allows for these meaningful discussions to take place in a collaborative way, and to ensure traditional trails like Lake One continue to be accessed with respect for and in reciprocity with the land. Only in this way will this trail knowledge be conserved and respected for all time.

NOTES

1 In this essay, the term "Aboriginal" is meant to include First Nations and Métis peoples and may be interchanged with the term "Indigenous peoples" as it is applied across Canada.

2 Robert Moor, *On Trails: An Exploration* (New York: Simon and Schuster Paperbacks, 2016), 2–3.

3 Pat Marcel, Member of Athabasca Chipewyan First Nation (Pat Marcel, Chipewyan elder, personal communication, 28 July 2014).

4 Susan Buggey, *An Approach to Aboriginal Cultural Landscapes*, Historic Sites and Monuments Board of Canada Framework Paper (Ottawa: Parks Canada Agency, March 1999).

5 Tim Ingold, ed., "Hunting and Gathering as Ways of Perceiving the Environment," in *The Perception of the Environment: Essays on Livelihood, Dwelling and Skill* (London: Routledge, 2000), 42.

6 Moor, *On Trails*, 2–3.

7 Parks Canada, "Wood Buffalo National Park of Canada, Park Management, World Heritage and International Significance," accessed 11 September, 2017, http://www.pc.gc.ca/en/pn-np/nt/woodbuffalo/info/plan.

8 United Nations Educational, Scientific and Cultural Organization (UNESCO), "Wood Buffalo National Park," *UNESCO World Heritage Convention* (1983), accessed 17 April, 2017, https://whc.unesco.org/en/list/256.

9 United Nations Educational, Scientific and Cultural Organization (UNESCO), "Wood Buffalo National Park."

10 The Lake One study area is contained within what was called the "new park," the lands annexed in 1926 to the original park lands. Between 1925 and 1928, over six thousand bison were shipped by rail to Waterways, Alberta, and from there by barge downriver to the park. During the winter of 1925–6, the plains bison migrated across the Peace River to feed on the lush meadows of the Lake Claire area outside of the park. Patricia A. McCormack, "How the (North) West Was Won: Development and Underdevelopment in the Fort Chipewyan Region" (PhD thesis, Department of Anthropology, University of Alberta, 1984).

11 J.D. Soper, "1932 Papers, Diaries and Notebooks," transcription from the Specimen Catalogues of J. Dewey Soper, listing the birds collected in Alberta (1918–1937), Accession PR1972.0294, item 0001. Box 1–7; J.D. Soper, "Faunal Investigations in WBNP 1932–1934," J. Dewey Soper Fonds 381. University of Alberta Archives. Accession no. 1979–21, (1934), accessed 18 February, 2015, https://discoverarchives.library.ualberta.ca/index.php/wood-buffalo-park-scene-2; Hugh M. Raup, "Range Conditions in the Wood Buffalo Park of Western Canada with Notes on the History of the Wood Bison," *Special Publication of the American Committee for International Wild Life Protection* 1, no. 2 (1933); Hugh M. Raup, "Botanical Investigations in Wood Buffalo Park. Canada Department of Mines," *Biological Series Bulletin* 74, no. 20 (Ottawa: Queen's Printer, 1935): 1–174.

12 A "cat trail" is a trail or road that was widened with a machine known as a caterpillar, or "cat."

13 Maxwell Graham, *Canada's Wild Buffalo: Observations in the Wood Buffalo Park 1922* (Ottawa: F.A. Ackland, King's Printer, 1923).

14 Many lakes in the national park were renamed by early explorers, and some were numbered and labelled this way on early historic maps (e.g., Lake One, Lake Two, etc.).

15 Dene language translation provided by Agnes Cheezie (personal communication, 23 September 2021).

16 Donalee Deck, *Lake One Trail Survey and Overview of the Archaeological Investigations at the Lake One Dune Site 2014*, Wood Buffalo National Park. Permit Report (Winnipeg: Parks Canada, January 2017); Marc Stevenson, "Scratching the Surface – Three Years of Archaeological Investigation in Wood Buffalo National Park, Alberta/N.W.T. – Preliminary Summary Report," *Research Bulletin*, no. 201 (Gatineau, QC: Parks Canada, 1983); Marc Stevenson, *Window on the Past: Archaeological Assessment of the Peace Point Site, Wood Buffalo National Park, Alberta.* (Ottawa: Canadian Heritage, 1986); Kaye Lamb, ed., *The Journals and Letters of Sir Alexander MacKenzie* (Cambridge: Published for the Hakluyt Society at the University Press, 1970).

17 Isidore Simpson (grandfather to Charlie Simpson) was born or baptized in 1889 (Pat McCormack, personal communication, 1 May 2024)

18 Group trapping areas were introduced in the park along with the park's new game regulations in November 1949.

19 Throughout this essay, the terms "cat road" and "cat trail" refer to the same type of trail.

20 A bison corral was built in 1965 at Lake One due to the abundance of bison in the area and was in operation for one year.

21 Patricia A. McCormack, Native studies professor at the University of Alberta and supervisor, personal communication, 24 May 2018.

22 The 1926 expansion of WBNP greatly impacted trapping and access to the park south of the Peace River. (See Patricia A. McCormack, "Chipewyans Turn Cree: Governmental and Structural Factors in Ethnic Processes," in *For Purposes of Dominion: Essays in Honour of Morris Zaslow*, ed. Kenneth S. Coates and William R. Morrison (North York, ON: Captus Press, 1989), 125–38; McCormack, *How the (North) West Was Won*). An interview with Archie Simpson was conducted for the "Fort Chipewyan Way of Life Study." He recalled his travels in the park growing up and raising a family at Peace Point. He was also a former WBNP employee.

23 A dovetail is a special kind of tight interlocking joint.

24 "Peace Point Log Cabin Condition Assessment," Memo to Laura Peterson, cultural resource management adviser, WBNP, from Kym Terry, Parks Canada Restoration Workshop (2000), correspondence on file, Winnipeg.

25 Another historic cabin is the Jackfish Patrol cabin, up the Peace River approximately twenty-five kilometres from the Simpson cabin.

26 A1 is a geological feature also associated with the spiritual significance of the Lake One Prairie.

27 Also known as "Archie's Trapline." This is another important overland trail that continues from Peace Point north to Robertson Lake in WBNP.

28 There are still descendants of the "Caribou Eaters" in the smaller communities around the northern area of the park, including in Łutsel K'e, NWT, and Fond-du-Lac, Saskatchewan (Sandra Dolan, local Historian and writer, Fort Smith, NWT, personal communication, 7 May 2024).

29 Patricia A. McCormack, Native studies professor at the University of Alberta and supervisor, personal communication, 24 May 2018.

30 House Lake was the location of an early Chipewyan settlement near Birch River on the south-western shore of Lake Claire in the Peace-Athabasca Delta. Many other families from House Lake also relocated to the Peace River, and eventually the House Lake settlement was abandoned.

31 Thomas D. Andrews and Susan Buggey, "Authenticity in Aboriginal Cultural Landscapes," *APT Bulletin* 39, no. 2/3 (2008): 64.

32 Tim Ingold, ed., "The Temporality of the Landscape," *World Archaeology* 25, no. 2 (1993): 152–74; Kent C. Ryden, *Mapping the Invisible Landscape: Folklore, Writing, and Sense of Place* (Iowa City: University of Iowa Press, 1993); Keith H. Basso, *Wisdom Sits in Places: Landscape and Language among the Western Apache* (Albuquerque: University of New Mexico Press, 1996a); Keith H. Basso, "Wisdom Sits in Places: Notes on a Western Apache Landscape," in *Senses of Place*, ed. Keith H. Basso and Steven Feld (Santa Fe: SAR Press, 1996b), 53–90; Vishvajit Pandya, "Movement and Space: Andamanese Cartography," *American Ethnologist* 17, no. 4 (1990): 775–97.

Chapter Seven

The Berry Picker

ATLANTA GRANT

Figure 7.1. The berry picker

Source: Courtesy of Atlanta Grant.

the berry picker.

pick.

small strawberry thumping in the palm of my hand,
I wrap my hands around my ribcage
spread my ribs wide open, place it inside.

thump.

my own little heartbeat.

thump.

keeping in time with the current of the river.

thump.

I breathe in the syrupy sweetness,
I am still here.
I place my hand on the soil.

thump.

pick the seeds out of my teeth,
take a deep breath.
dive headfirst into the water,
splash.

The poetry in this essay is a form of linguistic processing, beginning from the mind of a small wandering child who befriended the strawberry bushes in her grandparent's backyard, extending to a woman residing on the West Coast of Canada trying to make sense of her past while reconciling her futures.

It's a metaphoric representation of the identity politics and complexities that surround holding a multiracial identity within so-called Canada. A transparent, curious, playful, timid, muddled *process* of understanding my identity as a Métis woman with mixed German and Irish ancestry, and whose family histories and oral stories acknowledge Wendat and Métis origins as I reclaim what colonization attempted to erase.

My voice is an Indigenous voice, but it is a certain type of Indigenous voice. I am not a registered or status member of any nation; rather I have resided in urban environments for most of my life. With my voice in particular I feel called to uplift other Indigenous scholars, activists, and community builders – the hunters and the berry pickers. Bringing my own thoughts into an honest space that acknowledges how German I am, and how German I'm not. How Irish I am, and how Irish I'm not. And how Indigenous I am, and how Indigenous I'm not. I have been gifted the honour of walking alongside Indigenous people on the West Coast of so-called Canada throughout the past couple of years while completing my studies. And what has become evident is how important it is to honour the Indigenous knowledge(s) that have been shared with me, to learn how to hold them well. But additionally, how important it is to honour the stories and experiences that surround my mixed ancestry – to operate in an act of resistance, in *opposition* to the darkness that has been cast onto Indigenous bodies and minds through the violence of colonization. Personally, engaging in such resistance begins with the vulnerable tripping, falling, and stumbling into the settled ground of reclaiming my whole self, adding to the voices of urban mixed-ancestry people who are also walking a tightrope between identities.

It means recognizing where I should and shouldn't take up space. Where my voice should be shared, and where it shouldn't. Acknowledging the privilege I carry while continuing efforts towards unravelling how I am called to uplift and show up, while also being humbly transparent about the hardships and pain associated with trying to put together the puzzle pieces of my identity that colonization so freely scattered.

I am immensely grateful for the Indigenous peoples and communities who have shared their teachings and stories with me throughout the years, and I am humbled by their generosity of knowledge.

Foreword and Acknowledgments

Two stories are interwoven within the body of this chapter. The first, told through my own poetry and stories, follows a little girl who has berries for friends, as they teach her about her mixed ancestry and how to honour their teachings of reciprocity and gratitude. The second, making up the bulk of this chapter, addresses my academic journey as a graduate research student, which focused on contextualizing the "industrial food system" as presented within a North American context as an oppressive system that embodies and perpetuates social and environmental harm, as made evident throughout my ongoing research and interpersonal learnings around my own relations and habits with food, and as revealed through the Indigenous teachings and processes of food "cycling."[1]

Learning about and unravelling one's relationship with food[2] can be a parallel process of learning about where you come from, and how you are called to show up within those spaces. The following is a living piece as I puzzle through my own relationship to these issues, reflecting on my relationship with food and the food that I've wasted. Every letter, every word, represents an ebbing and flowing state of consciousness that will change as time goes on and is grounded in gratitude for the present.

I wish to acknowledge the land, place, and time this chapter was a part of to practise good relations towards the spaces I have lived/live in as a graduate student in British Columbia,[3] and towards the spaces I grew up in when I lived in Sudbury and Queensville, Ontario,[4] and to honour the stories shared and learned from throughout the body of this chapter.[5] I acknowledge this specific land/place/time framework out of respect to the earth and its natural environments, and those communities (such as First Nation communities within so-called Canada) who have operated from a space of equality, reciprocity, and harmony-focused relations with the earth since time immemorial and who are forced to fight for their land rights and sovereignty today. The following food-cycling practices were discovered through my own lived experiences, personal communications (where knowledge was shared by consent), and literature reviews via public domain documents. They hold complex nuances towards the land, place, and time frame for which they were originally practised (on-reserve, off-reserve, in urban environments, within the academy, etc.). Furthermore, it is important I address and acknowledge that many of the following food-cycling teachings are stories and recipes that I have gathered mainly from publicly available literature. It is very important for me to note that a lot of anthropological literature that holds Wendat teachings are written by non-Wendat people, most specifically white-settler anthropologists visiting community. This isn't an entirely accurate depiction of Wendat food-cycling teachings, since for a full representation it is vital to include community voices. This essay is the very beginning, an honest and vulnerable sharing of what my *initial step* looked like when learning about Indigenous culture and navigating my mixed ancestry.

These food-cycling practices are not just a "one-size fits all" approach that transcends the parameters of time, whether time immemorial, the 1800s, the 1900s, or the present day; rather, they have been/can be altered as colonial and capitalist agendas within the industrial food system continue to alter Indigenous people's cultural knowings and food-related ceremonial practices (such as hunting, fishing, medicine gathering, etc.). For land, place, and time, every Indigenous community is different, has unique traditions, and has faced its own specific experience of colonial histories and present-day colonial oppression.

Depending on the geographic location of a nation, colonialism has had different impacts on the survival of Indigenous knowledge(s) and one's cultural access to it. As my learning journey continues, I hope to always maintain the good relations of acknowledging the systemic, oppressive barriers in place, and this land/place/time approach to food-cycling knowledge, biocultural heritage, and other Indigenous knowledges I learn from. While Indigenous peoples and communities may share commonalities, there are many differences. Contextualizing the Indigenous knowledge(s) that are about to be shared is necessary to understand the complex histories and present-day colonial agendas Indigenous peoples face in regard to knowledge preservation today.

schools in session.

kneeling down
bare kneed on the front porch,
lines dig into the crevices of my skin.
It smells like sickly sweet strawberry air,
as the August heat turns their insides to jam.
I turn my ear to the south.
cicadas.
schools in session.

Mapping and Understanding Indigenous Knowledge(s)

As my learning journey continued, I found myself wondering how "Indigenous knowledge" was defined. What did it encapsulate? Here, I embarked on research around the many categorizations of Indigenous knowledge (most specifically, in the realm of ecology/environmental and food-related studies), how it is understood, broken down, and partnered with in consented cross-cultural spaces (Indigenous and non-Indigenous for example). This was not only to ensure its protection and culturally appropriate use within my own work, but also to personally unravel and begin to understand what type of relationship I have *with* and *towards* its teachings.

As Anishinaabe scholar Deborah McGregor states, "despite the interest in TEK [traditional ecological knowledge] by environmental managers, policy makers, academics, consultants, environmentalists, and Aboriginal communities themselves, the meaning of TEK remains both elusive and controversial."[6] TEK is a commonly used categorization of Indigenous knowledge (as it pertains to the environmental and food-based literature). Many Indigenous scholars who have broken down its use and understandings, such as Potawatomi philosopher and climate/environmental justice scholar Kyle Whyte, discuss TEK and its role in ecological knowledge as a collaborative concept, focusing less on definitions and categorizations and more on how these different epistemologies allow for varying approaches to environmental stewardship. In cross-cultural and cross-situational collaboration across Indigenous and non-Indigenous institutions, "any understanding of the meaning of TEK is acceptable only so long as it plays the role of bringing different people working for different institutions closer to a degree of mutual respect for one another's sources of knowledge."[7] This definition of TEK means environmental scientists and policy professionals, Indigenous and non-Indigenous, should focus more on creating long term processes that allow for the implications of different approaches to knowledge in relation to stewardship and management priorities to be responsibly thought through.

To me, this emphasizes that it is less important to focus on definitions about what Indigenous knowledge *is*, and more important to focus on what long-term processes can be created so that Indigenous knowledge(s), when shared within cross-cultural collaborative spaces, can be safely interacted with – thereby enforcing mutually respectful knowledge exchanges between differing world views. Here it becomes not just knowledge *about* relations and land, but a *participation* in cross-cultural collaboration for the reciprocal stewardship of the land.

Other categorizations include "biocultural heritage," the knowledge and practices of Indigenous peoples and their biological resources that are "held collectively, sustains local economies, and are transmitted from one generation to the next. [This] includes thousands of traditional crop and livestock varieties, medicinal plants, wild foods, and wild crop relatives. These precious resources have been conserved, domesticated, and improved by communities over generations – and sometimes millennia."[8] This definition provides powerful context and consideration for a movement away from the concept and language of "traditional" when describing Indigenous ecological/environmental knowledge. And while the concept of biocultural heritage still highlights various practices, crop, and food varieties as "traditional," it doesn't situate Indigenous knowledge as a "traditional" or "historical" entity, which would imply that such knowledge is outdated. Rather it highlights practices that in their nature have been used traditionally over time.

Lastly, "Native science," as discussed by Gregory Cajete, is described as a "process for understanding in all aspects of Native tradition … [It deals with the] systems of relationships and their application to the life of the community."[9] An integrative outlook involving tools that encourage meaning and understanding versus prediction and control, Native science is the process of learning about our relations with and towards the land and the responsibilities attached to entering into relationship with it.[10] Within this understanding, Indigenous peoples seek to learn from their relationships with the Land, holding its teachings harmoniously to be shared with the wider community, with emphasis on the intergenerational transmission of such teachings not solely for community benefit, but for reciprocal relations between the earth and all of its inhabitants, further demonstrating that knowledge is not merely for the individual but something that should be shared for collective and collaborative benefit.

As my learning is ongoing, and I unravel various understandings of Indigenous knowledge, I currently choose to use the term "Indigenous biocultural heritage," as opposed to "traditional ecological knowledge" (for example), to engage in an action-oriented step away from the concept of "traditional." "Traditional," as it is commonly associated with "outdated," "past," and "historical," potentially places labels on the knowledge and practices Indigenous people hold, which may continue to situate Indigenous knowledge(s) in the past instead of acknowledging their fluidity and significance in present-day environmental discussions.

My desire is to understand how to safely enter into cross-cultural relationships between Indigenous and non-Indigenous communities, to understand what entering into a system of responsibility with Indigenous knowledge(s) entails.[11] Here, Whyte suggests avoiding placing the "supplementary value" label on Indigenous knowledge.[12] Meaning that if any Indigenous teachings are discussed and/or utilized in non-Indigenous spaces (when permitted), they should not be discussed as a *supplemental* thing, in *addition to* a more "dominant" or "credible" form of knowledge. In other words, they should not be subsumed within dominant Western individual knowledge and epistemologies, in ways that regard the addition of Indigenous knowledge as merely helpful or interesting versus credible, valid, with intergenerational first-hand experience of environmental stewardship and governance. Through such avoidance, perhaps we may see such knowledge exchanges occur in ways that uplift Indigenous voices, and support Indigenous governance, autonomy, and sovereignty.[13] However, Whyte reminds us to proceed with caution, as "TEK cannot be readily transferred to different contexts unless the people in the new context also learn the systems of responsibilities and character traits."[14] This brings additional questions to light: Do those wishing to learn from Indigenous knowledge systems merely want to fit such teachings into an already established policy framework (for example), or do they seek to change the framework itself? Do those engaging with Indigenous knowledge seek integration for integration's sake, or do those parties *also* seek to enter into a system of responsibility, which means engaging in a *process of understanding* around how such differing world views and knowledge(s) will work together (or apart) when it comes to such initiatives? By viewing entering into relations with Indigenous knowledge(s) as a collaborative practice, perhaps we can avoid this integration narrative. Through understanding Indigenous knowledge(s) as collaborative *systems of responsibilities* that hold teachings towards mutual respect and reciprocity, and by avoiding the submissive undertone that often accompanies efforts to include Indigenous knowledge(s), there may be cross-cultural relationships and partnerships in which Indigenous peoples control their own knowledge(s),[15] and share such knowledge(s) in a way that ensure they are respected and understood as valuable in their own terms, and not only as a mere *addition* to pre-existing non-Indigenous environmental systems.

This requires what philosopher Stolomowski discusses within the context of the "eco-person": "one who perceives ... [that] lasting solutions cannot be instant solutions. The first step in the work is one of inner reconstruction, so that we achieve some balance, some harmony within, some clarity of vision, the sense of our place in the larger universe; that we acquire in short, wisdom."[16] Personally, I have found great rooting in Deborah McGregor's discussion of the Two Row Wampum as an ideological framework for cross-cultural collaboration to accomplish this.[17] The Two Row Wampum and its model of coexistence is founded on the belief that having separate world views are not necessarily an undesirable thing, and that developing a framework that would respect different world views would be an appropriate approach to take.[18] This values both systems, where, for Indigenous peoples, their cultural survival becomes assured, protected, and legitimized – their role in environmental leadership recognized and upheld.[19] "This is a long-term, mutually beneficial relationship, where each side respects the other's worldview and their right to live accordingly."[20] The Two Row Wampum serves as a historical and present teaching of what coexisting looks like, and how two different belief systems are meant to be a strength and not a disservice to one another. This approach also illustrates how this model of governance will serve Indigenous communities more strongly than present integration and assimilation models that currently do not seek to practise or understand the responsibilities attached to entering into a relationship with Indigenous knowledge systems.

Seed Burier, Berry Picker

Growing up, there were certain rules and undeniable truths I held in my relation to the world: (1) be gentle when walking through the woods; (2) never take more food than you can eat; and (3) the strawberries, if you listen and look *really* closely, are our most receptive friends. I carried an interconnected world view that practised relational accountability towards the earth.[21] Of course, as a child, I did not think of it as such; these simply, were special generational teachings and observations held by a little girl who had a backyard full of berries who had nothing else to do but listen intently to her musings. As a mixed German and Irish woman, my world view has led to practices that involve considerations towards the food that I consume, and the food that I "waste." However, as I transitioned to urban life, I found it difficult to protect my food and began disposing of it in a way that did not show equal care towards the environment I once carried such relational accountability towards. Personally, my childhood experiences have illuminated the distinct role that beliefs, ideologies, and world views play in shaping our relationship to the natural environment, plants and the living creatures residing here with us. I have learned that the loss of one's relationship with the berries as a companion and friend distorts our relationship with the natural environment, and causes us to forget to care and provide stewardship.

As I began my graduate studies on the topic of Indigenous food sovereignty, I began to learn how unique every Indigenous community's biocultural heritage is and the ways these are connected to my own desires to learn about how to mend my relations with food. These are teachings that call us to be intentional with what we consume for food and how we enter into relations with it. Most illuminating to me, however, was how through such teachings it came to seem backwards to even consider food "waste" as something categorical or something that can be imagined or possible within such relations. It seemed backwards, then, to "waste" my food, as food was never meant to be wasted.

seed picker.

we are here now.
I come to you learning.
wrapping every inch of my vertebrae around every stalk of corn,
sisters, I am here now.
standing tall
you see,

you tried to bury me,
but
me?
the little girl who had berries for friends?
berry picker?
well,
I've been swallowing seeds my whole life,
pickin' em right outta my teeth.
and
sowing em'
right back into the ground.
and just watch,
I'm sprouting right back up.
blooming,
right.
into.
the sun.

Over the last several decades, major changes have occurred to food systems in the Global North.[22] This can be witnessed through the expansion and movement away from small-scale, land-dependent (i.e., seasonally shaped, local) systems to a dis-embedded industrial system, both of which are entangled with increasingly globalized political and economic systems. This movement towards a global, trade-based, and now industrial system is made possible through the introduction and emerging reliance on agricultural technology, or "agro-technology," to aid in and ensure global markets and capitalist demand can be met. While agrotechnology is one example among many that can be linked to the emergence and growth of the industrial food chain and system, it is also a powerful example of how such technology and "industrialization" encouraged humans to maintain a certain distance from the land. Moving away from a more "hands-on" relationship allowed for an "extractive-based" mindset – fuelled by demand for large surpluses, quick timelines, and economic performance – to become the dominant link between humans and what we consume for food. Such relations have created a greedy and unequal relationship between humans, the natural environment, and the diversity of cultural knowledge systems that grow from it.

As global populations expand, the demand for continual growth becomes an expected trajectory stemming from commodified capitalism, intensifying the pressure for mass-produced food and the infrastructure to support it. Within the North American context, this allows for trade of products globally and a market demand where middle- and high-income communities have access to a variety of foods, whereas marginalized communities – those situated in a system of food apartheid[23] and subsisting on lower and middle incomes – do not experience this food system or have access to food in the same way.[24] This system spreads inequality and food injustice towards those engaging with it, injustices that can be seen via the unequal experiences felt across food access and food survivance[25] movements. One particular injustice, however, can be exemplified through the produced output of food waste.

The discourse on food waste emerged as a subcategory of "food security" beginning in the 1950s as food production increased dramatically due to increased "efficiency" within food chains. However, at the same time, many communities still did not have access to food and were relying on food banks and relief efforts.[26] At this time the verbiage began to shift from "Why is there so much food leftover?" to "Why is there so much waste?"[27]

Food waste is not accidental; it actually feels rather *intentional* when we begin to understand its historical origins. However, as I am reminded by my Indigenous colleagues and friends, the origins of food "waste" can be more accurately located within the histories of colonization, originating through the enforced, extractive, violent, and careless relations exhibited towards the land and its people, as seen with the genocide of Indigenous peoples and their knowledge systems. Waste is not imagined; it is designed by the colonial systems in place.

Food "cycling," however, as will be discussed below, is an intentional choice to not waste. It is an Indigenized term that one may view as a decolonized adaptation of food "waste," but which in fact refers to the processes and teachings that Indigenous people since time immemorial have been practising in relationship with the land. And while such practices were disrupted through the violence of colonization, they continue to be practised and revitalized today. Food "cycling," as will be shown, holds critical knowledge and practices that can help combat broken relationships with the land, support future utilization of food pathways, and, most importantly, uplift and provide spaces for Indigenous governance and autonomy to occur. Despite decades of colonial violence in relation to Indigenous food sovereignty and food survivance, Indigenous peoples continually speak up for their inter-species relatives on this land. Most, if not all, Indigenous oral traditions convey teachings regarding food handling that are carried out with respect for their animal and plant relatives, who provide sustenance for their communities and selves, a selfless gift that is deserving of intentional honour and intentional repurposing.

For the remainder of this journey, I will now explore what I learned about these relationships through the teachings of the Wendat peoples

in the realm of biocultural heritage and food "cycling." I discuss Indigenous teachings from a specific nation in this essay to ground Indigenous
knowledge in a specific place and set of relationships, and have chosen
Wendat teachings to honour my mixed ancestry and family histories,
which contain oral stories that acknowledge Wendat and Métis origins.

fresh snow hit the ground.

I pretend I know where all the bears are hibernating in the woods behind
gramma's house.
all my berry friends went to sleep for the winter.
"don't step where they grow."
gramma's voice in my head.
I don't remember the way back to the trail,
see my dog, mac, running full speed ahead.
I put my feet right in his paw print.
big paw in little paw.
I follow.
crunch.
no one would even know I was here.

Huron-Wendat First Nation: Nionwentsïo and the Sacred Circle

Within these early stages of learning, establishing the following food-
cycling teachings within the Lands from which they originated is very
important. I felt it best to introduce the geographical location of the
Wendat peoples ("Huron" being an imposed name given by the French)[28]
within their own words. As the Huron-Wendat Nation states,

> The Huron-Wendat Nation has approximately 4,000 members and almost
> all of them are French-speaking, which makes it the only francophone com-
> munity in the Iroquoian linguistic family. Almost all of the Huron-Wendat
> members reside in Wendake, the only Huron-Wendat reserve in Canada.
> According to the Huron-Wendat oral tradition, the customary territory of
> the Huron-Wendat Nation, the Nionwentsïo, which means "our magnifi-
> cent land" in the Huron-Wendat language, lies along the north and south
> shores of the St. Lawrence River, between the Saint-Maurice River and the
> Saguenay River. This is very special territory for the Huron-Wendat and is
> described as an exceptional hunting and fishing area, a major travel axis,
> as well as a centre of trade and diplomatic relationships, full of abundant
> resources and infinite wealth. The members of our Nation have lived
> and moved about there since time immemorial. The St. Lawrence River,

the Great River according to the Huron-Wendat oral tradition, which is situated at the very core of the Nionwentsïo, is the highway used by the Huron-Wendat since time immemorial to reach the territories where they traditionally practised and continue to practise their traditional activities of hunting, fishing, and trade. The Great River thus occupies a central place in the identity and culture of our Nation, who is its guardian and protector.[29]

The Sacred Circle and the Introduction of Food Cycling

As my learning continued, I discovered Wendat scholar George Sioui's book *Huron-Wendat: The Heritage of the Circle* (1999), which discusses the sacred circle of relationships that are held between all beings on earth as participation in an interwoven kinship. As humans, we can either acknowledge our role together as one within the circle, or the circle itself will be forgotten along with the network of kinship it holds.[30] Sioui encourages us to acknowledge reciprocal relationships between humans and non-humans, and to not dismiss them as being lower in the so-called hierarchy of life. He argues that the sacred circle of life, in which the place of humans is equal to that of the other creatures, albeit marked by a special responsibility, is divided into four quarters. When they have covered the entire circle, they speak the words "all my relations," thus acknowledging the relationship between all beings in the universe and their common vision of peace.[31] Humbly learning from George Sioui's book and its Wendat teachings alongside the teachings that have been gifted to me throughout my time on the West Coast, I am reminded of the importance of cyclical relations. How "waste" as it is now defined and understood originated from an oppressive and violent industrial food system, and cannot continue to be used if we are to see futures of mended relations between humans and what we consume for food. I see such cyclical similarities in conversations I had with members of Kitasoo Xaixais First Nation throughout my graduate studies within the language reframing[32] of food "waste" into food "cycling."[33] As Gregory Cajete notes, "Language is more than a code; it is a way of participating with each other and the natural world. At one level, language is a symbolic code for representing the world that we perceive with our senses. At a deeper psychological level, language is sensuous, evocative, filled with emotion, meaning and spirit...language is animate and animating...in the Native perspective, language exemplifies our communion with nature rather than our separation from it"[34]

Food cycling is the act of participating in harmonious relations with what is consumed for food through acts of intentional care. Food cycling is the journey between what is consumed as food, from harvest and hunting to consumption, and any other process that occurs after consumption. Food cycling, when it can be culturally practised (in the face of threats from environmental degradation and knowledge exploitation), is the spirit of intentional care that ensures the living – animal, plant, and all other beings that come and go from the Land – are honoured to their fullest capacity. It is the resounding example of Indigenous intergenerational transmission of knowledge and has traditional significance and present-day adaptations that mirror Indigenous adaptive capacity and the ongoing cultural revitalization that exists today inside and outside community. Within my graduate research it became clear that food cycling follows two processes: (1) intentional honour, and (2) intentional repurpose. Here, the process of intentional honour is exemplified when an animal, plant, or other gift from the Land is honoured before consumption through a hunting or harvesting practice, and during consumption through ceremony or offering, and after consumption via ceremony and burying. In other words, it occurs at every stage of the food cycle towards what is consumed for food. Intentional re-purpose highlights the after-consumption stage, where the animal, plant, or other gift from the Land is transformed into something else worthy of intentional honour, which is repurposed, transformed, and ignites the start of a new cycle and journey as cultural regalia or medicine, for example. These practices and their associated processes (e.g., harvesting herring eggs, burying salmon bones in the ocean floor, utilizing all parts of the animal) are all interwoven and cyclically a part of the network of responsibilities and practices that are attached to Indigenous biocultural heritage as a whole.

The Animal Life Cycle: Relational Teachings and Practices

Animal-centred practices, such as hunting, fishing, drying, smoking, and additional ceremonial practices, are very time-specific, land-specific, and complex. Stemming from Wendat histories, land-based hunting and agriculture would vary depending on where one resided. If we are looking through a historical lens of literature, within the scope of Ontario and Quebec geographies the literature suggests that for the Wendat, fishing and land-based hunting encapsulate the majority of agricultural and land-based activities. As noted by Birch and Williamson, "The hunting of white-tailed deer (Odocoileus virginianus) and wapiti (Cervus canadensis) by the site inhabitants for hides, meat, and bone for making tools, therefore, must have been one of the most important economic considerations and planned seasonal activities."[35] Here, the animal is fully utilized, including its brain, liver, pancreas. Food parts are prepared via boiling, roasting, and drying techniques,[36] and the reserved water from the cooking process is used as broth for soups. Leftover fat and/or tallow from the animals (deer, bear, etc.) can be kept and used as medicine.[37] Here, material (moccasins, packaging, etc.), rattles, and other instruments are made from skin and fur, and sinew (animal tissue) used at times to thread, with these actions of intentional repurpose providing humans with food, clothes, shelter, and direction.

Fish, such as trout, historically and in the present, are caught and cooked, which includes smoking and preservation techniques – with bone burial (and other utilizations of the fish bone) used in ceremony.[38] Here, remaining bones are placed back into the river/ocean/land for reciprocity and cyclical teachings, which presently are being connected to a plethora of positive environmental benefits within the academic and quantitative research sphere. For example, a research study done in the State of Washington has shown that the addition of salmon carcasses (bones and leftovers) to rivers positively correlates to stream productivity and nutrient levels.[39] This life-cycle practice concludes the utilization of the fish in its entirety, with future fishing and harvesting decisions based on the number of salmon that return for the protection of salmon conservation and well-being. As well, this is exemplified in death, as salmon continue to provide sustenance for other creatures, with their physical decomposition providing positive environmental benefits to the land.[40]

Overall, Indigenous ontological practices share unique and distinct life-cycle-handling practices, such as whole utilization of the animal, repurposing of bones, organs, and hide, and the careful placement of leftover pieces for decomposition. Each of these practices are centred around an ontology in which animals are viewed as spiritual beings with whom humans are engaging in a non-hierarchical interconnected relationship, learning from one another and showing mutual reciprocity and responsibility.[41] Every decision is meticulous and made with careful reason rooted in each specific community's traditions and oral generational histories and stories, showcasing, "a long-term relationship of reciprocal exchange between animals and the humans who hunt them."[42] While forces of colonial erasure have ensured some Indigenous communities have lost these teachings, cultural revitalization and intergenerational story-telling proceeds onwards to combat continued erasure, with other Indigenous communities practising these teachings freely today.

honey.

she often found herself hiding between the house and entrance to the forest.
strawberry plants, its fearless guardian, create a makeshift fence
as if to say,

only a select few are welcome here.
you see,
she wasn't trying to intrude;
she thought this was a good place to hide.
tried picking the seeds out of her strawberry plants
ended up turning them to jam.
"*I always push too hard.*"
lost.
she ran backwards to the pond.
hands sticky
remnants of the life she took,
she tries rinsing her hands in the cool water.
sunlight reflecting off the corner of her cheekbones.
sticky
sweet
red memories floating away with the reflection.
she licks the rest off her hand.
"you know you are what you eat."
beaver popped his head up above the dam.
"well I'm just trying to be sweeter."
bitter**sweet.**
that lesson was.

Intentional Repurpose Part 1: Listen to the Plants and They'll Tell You

While animals fill a certain need, the rest of nutrition and sustenance is often gathered in nature by hand. Considering the geographical location of the Wendat, a notable agricultural practice is exemplified through the creation story of the Three Sisters, a polyculture farming technique involving the intercropping of corn, beans, and squash.[43] This included land-specific irrigation techniques and practices that ensure the survival of these agricultural crops. Here all seeds are planted within the same mound, where "corn provides tall stalks for the beans to climb so that they are not out-competed by sprawling squash vines. Beans provide nitrogen to fertilize the soil … The large leaves of squash plants shade the ground which helps retain soil moisture and prevent weeds."[44]

The planting of the Three Sisters showcases two things regarding food cycling. Firstly, through current domestic forms of gardening and planting, we engage in assumptions about our physically constructed environment that a human-made space will work for the plant.[45] Secondly, it shows us a less wasteful and more relational practice of growing food.

Present-day literature speaks to the practice of intercropping as less susceptible to pests, meaning there is less need for pesticides and fertilizers and less need to use water.[46] While this does not speak to the physical food wasted, it speaks to additional forms of waste within the food system, such as water waste and the associated negative impacts. These practices help us reframe our view of plants and foodstuffs as singular-use objects to multi-use life forms of medicinal healing in waste prevention.

In contrast, within forms of monoculture (growing one crop at a time) there is heavy reliance on technology inputs, such as machinery, labour, and fertilizer usage.[47] Due to single cropping, there is diminished engagement with traditional skills surrounding the gathering and growing of crops. The planting of the Three Sisters showcases lessons of reciprocity and relationship in growing food, and indicates how through this relational lens, we indirectly contribute to food-cycling practices that are less wasteful towards the land. Current cropping methods can strip us of our emotional connection to crops, as a single-minded focus on yield can translate into a loss of the relational values and reciprocal exchange associated with food. Without meaning, these crops/plants and foods are more easily wasted and produced for mass consumption as a commodity.

Additionally, maize (corn), as is historically noted within the literature, made up half of the Wendat diet,[48] and today is still a food product held in high ceremonial regard and is incorporated into many dishes. Historically, the planting of corn has been documented in community, as planted "when the blossoms of the juneberry appear."[49] This is a small but notable sign of the land dictating when it is appropriate to grow. While in large-scale industrial agriculture this may seem small, it indicates a time in which one observed and listened to the Land as teacher instead of telling the land how it will be used.

it just all feels bigger than me.

robin sits on the birdfeeder
stares right at me.
my nose pressed up on the screen door
saltine cracker crumbs sticking to my sweater.
turn around,
grandma's gone.
tried catching the robin with my bare hands.
I always wanted a bird as a best friend,
maybe if it felt my pulse
It'd know,
you are safe with me.

Corn was/is a heavily utilized multipurpose crop, with the husks used for mat making and basket weaving, and leftover (dropped, uneaten) corn used for seeds in the next crop cycle. There are many recipes including corn (these will be discussed below) that are broken down into cycles of ripeness for the crop, ranging from corn breads to soups. For example, "early bread is a recipe that utilizes 'newly ripened' and 'undried' corn which is made into a paste as the basis for the baked bread."[50]

Understanding plants' multi-use life cycles, eating foods based on what the land offers, and using foods in ways that take their life cycle stages into consideration (e.g., ripe, unripe) is less wasteful. This showcases mutual respect for the land for which the plant was gifted. We must acknowledge that foodstuffs aren't one-purpose foods; they have an entire life cycle consisting of different stages that deserves to be honoured through full usage and/or consumption.

Intentional Repurpose Part 2: The Sharing of "Recipes"

The following recipes highlight various understandings, languages, and utilizations of food that showcase a food-cycle approach. Recipes feel quite structural, containing certain requirements at times (measurements, specific amounts, etc.). Finding a word that better embodies the shared knowledge around food discussed within ways of feasting and food preparation – for example, the smoking and drying of fish, and/ or the drying of seaweed or meat – could perhaps better encompass this relationality to food. Nonetheless, through these "recipes" we can learn of a kind of shared knowledge surrounding food, and the utilization of foodstuffs that are commonly considered "waste" or "unworthy" for common consumption.

A notable theme within these recipes, as observed throughout the literature, is that they call attention to the various life stages of the food being used. Whether pumpkin, corn, squash, or various other crops, there is a recipe for its early life stage (not ripe) or peak life stage (ripe). This is not uncommon in the food waste movement rooted in our industrial food system and local farm-to-table movements; however, this is commonly discussed in the context of needing to utilize our *leftovers*, or of determining *how to repurpose your leftover foods*. Whereas here I'm highlighting a mindset in which the food is not *left over* but is instead viewed as *intentionally repurposed*. This is not to suggest that these recipes and food-cycling practices weren't potentially born out of an observation made around communities' leftover foods. But, considered in the context of Indigenous knowledge systems, there is always the acknowledgment of the Land as a giver, teacher, and provider, and one can quickly understand the difference in intention and language between honouring the life cycle of a crop versus repurposing *leftover* foods. Food isn't complacent; it is an active agent, a teacher, a friend. What we as human beings consume as food, and how we intentionally and relationally engage with it, can become a catalyst for how we treat one another in our communities and in relationship with the planet as a whole.

Courage My Love: Reflections and Contextualization(s).

she found herself back in the woods once again, its fearless protector, friend. seeking permission from the strawberry guardians, tired from trying to figure out everything at once. tired from running, popping blueberries between her fingertips.

gramma said not to take so many at once, slow down. calluses on her hands, little mountains on each knuckle, reminiscent of every hurtle she's faced in her life. running down to the river, choking on air, climbing on top of the highest rock, "I can feel something here."

arms out, chest puffed, pretending to be a chickadee bird. thought about flying. shoving pebbles into the corner of her cheeks, "chickadee deee deee deee." nothing echoed back. frustrated, she hops back to gramma's house, blueberry pie wafting through the wind, it carries her feet there.
see? I will hold you up.

shovelling in spoonfuls of pie, bite after bite. hot blueberries bursting in her mouth, she placed her feet down firmly underneath her gramma's table, roots wrapping around every vertebra, clicking her bones into place. "why do you feel the need to run from me? you got everything you need here." deep breath. blueberry pie. glass of milk. listening to the wind howl, it was all she needed to hear.

Food waste as a proposed issue seems to be wrapped up in many different areas of enquiry; there is not just one issue it can be presented as. However, do the ways in which current food waste researchers present and organize their "data" (i.e., through statistical measures of waste) allow for a threshold of waste to exist and be maintained? Does creating a narrative around food "waste" ensure that "waste" exists and becomes a "natural property" of our food that is wasted? Current food "waste" literature and knowledge continue to frame it as an "environmental issue," connecting food waste as a contributor to climate change[51] with specific

examples stemming from its correlation to the rise in greenhouse gases –
noting emissions at the individual and provincial levels,[52] an "economic
issue," describing food waste as economic mishandling resulting in loss
of federal and provincial capital,[53] a "surplus issue," or an "educational
issue" in which a lack of food literacy is noted as being potentially responsible for the disconnect between humans and our food and is presented
as a potential area of intervention to help mitigate food waste.[54] This
maintains a narrative around our food as an entity separate from us, as a
non-relational resource to waste. Supporting current injustices that stem
from the industrial food system, food waste and our current understanding of food waste are all consequences of such a non-relational system.

Most critically, the food waste literature and practices pose a serious
threat towards Indigenous advocacy. As land stewards, Indigenous activists are advocating for rebuilding broken Indigenous food systems with
place-based, relational knowledge that addresses the complexities of
food inequality, reminding us to reconnect and learn a valuable lesson
from the land. The fight for legal rights to access and practise culture on
the land, where the processes of food-cycling knowledge originated, and
have been shared for generations, can only continue if reconcili(action)
is taken, and spaces of reciprocity and collaboration are devoted to the
uplifting of Indigenous communities and Indigenous knowledge systems to support the reinstatement of Indigenous laws and knowledge
systems. Only then, can discussions of Indigenous biocultural heritage
and its role within the current settler-colonial industrial food system be
approached.

It matters deeply how food-related issues are contextualized. It matters
deeply how one frames a problem and its solution, and it matters deeply
how we contextualize the industrial food system, as how such framing is
presented can either cultivate change, meaningful change, or continue
to contribute to "change" that maintains the status quo.

It is important to note that the food-cycling practices and teachings
shared here are not to be understood as a "solution" to the crises resulting from our current industrial food system; rather, we should see them
as a step towards igniting conversations that enable behavioural changes
for the sake of our planet and relations with one another. I also hope to
share vulnerably how unravelling my own behaviours within myself led
to a beautiful journey around mending my relationship with my food.
These teachings I hold carefully, not to share as a solution, or as a form
of tokenism, but as worthy and valuable in their own right, not only when
attached to problems and/or reconciliation initiatives.

We can take lessons on how to best root our contrasting worlds, puff-
ing out our chests courageously, though, ultimately, work involving an

inner reconstruction of the heart is hard. It cannot be as simple as changing the way you see the world one day. It requires what Fletcher[55] and Scharper[56] suggest is a "structural uprooting," in which we don't merely embrace this knowledge and then continue to tend to the current capitalist structures we are working within,[57] but rather engage in systems of responsibility attached to the use of biocultural heritage[58] in the context of land stewardship today. There is no room for relationality, or listening to the environment in capitalist, objectivist, and economic structures, nor are there rooted capabilities in such structures for understanding the strawberry as a gift giver or the berries as our friends. Moving away from Indigenous knowledge as a cultural construction into literal and metaphorical truths,[59] and ensuring these teachings and forms of sacred knowledge synonymously serve the Indigenous communities in which such knowledge originates, are broad steps towards moving forward.

I am grateful to learn about Indigenous relational systems as I learn more about specific practices from Indigenous peoples, such as the Wendat described in this essay. Seeking to be respectful of lifeways that emerge from the land and its diversity of community is a lifelong journey, one that I approach with humility, hope, and encouragement. Ultimately, it is a winding road of unanswered questions and wrong turns, and requires the trust to breathe deep and fill your lungs with the confidence to ask questions and to start unravelling how you are called to show up today as a steward of this planet, and as a human being to one another.

the berry **giver.**

sometimes when the dust starts to settle it shakes up remnants of what once *was*.
light shining through the particles, you get a glimpse of what is *to come*.
whole parts of you settling into their new home.
I'm learning to like the in-between.
it's the final stages of nostalgia before letting the grief go and allowing an ending to rest well.
little me nods goodbye as I close the door behind me.
couldn't help but notice she left something on the front porch.
a strawberry seed.
I guess she knew,
even as we'd get older.
we would need to be reminded,
of where we came from.

....

It's trust.
It's the dust settling into its new home.
It's the picking up of courage.
courage, my love.
you
are courageous.
and you are right where you need to be.
trust me,
the berries told me so.

NOTES

1 Food "cycling" is a term that will be used throughout this essay. It has
been used in circular economy narratives, to describe the food nutrient
cycle occurring in a particular ecosystem, and in reference to individual
consumers' food waste journeys. Food "cycling" is an Indigenized term
that arose throughout my graduate studies and fieldwork alongside
community members from Kitasoo Xai'xais First Nation. This term operates
in opposition to the colonial term and action of food "waste" and contains
two processes of "intentional honour" and "intentional re-purpose."
It can be categorized as a part of the Indigenous knowledge system of
"Indigenous biocultural heritage," as will be discussed. On circular economy
narratives, see T. Lu and A. Halog, "Towards Better Life Cycle Assessment
and Circular Economy: On Recent Studies on Interrelationships Among
Environmental Sustainability, Food Systems and Diet," *International Journal
of Sustainable Development & World Ecology* 27, no. 6 (2020): 515–23. On food
nutrient cycles, see A. Keys, E.H. Christensen, and A. Krogh, "The Organic
Metabolism of Sea-Water with Special Reference to the Ultimate Food Cycle
in the Sea," *Journal of the Marine Biological Association of the United Kingdom* 20,
no. 2 (1935): 181–96. And for food waste journeys, see M. Setti, F. Banchelli,
L. Falasconi, A. Segrè, and M. Vittuari, "Consumers' Food Cycle and
Household Waste: When Behaviors Matter," *Journal of Cleaner Production* 185
(2018): 694–706.
2 "Food" within this essay refers to whatever humans consume for sustenance,
energy, nutritional value, and survival. What falls under this category is
different for every individual depending on their socio-economic status,
cultural background, and geographical location. It simply cannot be
categorized into one thing; it is a beautiful gift that has been picked
apart and calculated to look a certain way, taste a certain way, and be
accessible in ways that suit capitalist colonial frameworks of food access. As
Haudenosaunee scholar Adrienne Lickers Xavier notes, "In order to discuss
food waste, we need to unpack the paradigm that has categorized food as a

commodity." See T. Soma, B. Li, A.L. Xavier, S. Geobey, and R.F. Gutierrez, "All My Relations: Applying Social Innovation and Indigenous Methodology to Challenge the Paradigm of Food Waste," in *Routledge Handbook of Food Waste*, ed. C. Reynolds, T. Soma, C. Spring, and J. Lazell (London: Earthscan, 2020), 319.

3 I acknowledge the unceded and ancestral land the University of British Columbia resides on, and where I currently live, in so-called Kitsilano, Vancouver, British Columbia. These lands are the unceded traditional territories of the xwməθkwəy̓əm (Musqueam), Skwxwú7mesh (Squamish), and Səl̓íl̓wətaʔ/Selilwitulh (Tsleil-Waututh) peoples.

4 Moving around a lot as a child meant that a variety of different towns and cities housed my growth. However, I wish to primarily acknowledge the city of Sudbury, Ontario, as I was born there and it was my constant home, with my grandparents still residing in the home I spent most of my childhood in. I acknowledge their home resides within Robinson Huron Treaty territory, on the traditional territory of the Atikameksheng Anishnawbek. And I additionally acknowledge where I spent my teenage years – Queensville, Ontario, on the traditional territories of the Mississaugas of the Credit, the Anishnabeg, the Ojibwe, the Haudenosaunee, and the Wendat peoples.

5 I also wish to acknowledge that pieces of this essay have been published in my master's thesis, which is held at the University of British Columbia. Some of this material was also revised and included in *FOAM* Magazine's "FOOD! The Nourishing Issue." See A. Grant, "The Berry Picker," *FOAM International Photography Magazine*, no. 63 (February 2023): 225–32.

6 D. McGregor, "Traditional Ecological Knowledge and Sustainable Development: Towards Coexistence," in *In the Way of Development: Indigenous Peoples, Life Projects, and Globalization*, ed. Mario Blaser, Harvey A. Feit, and Glenn McRae (London: Zed Books, 2004), 77.

7 K.P. Whyte, "On the Role of Traditional Ecological Knowledge as a Collaborative Concept: A Philosophical Study," *Ecological Processes* 2, no. 1 (2013): 2.

8 See "Biocultural Heritage," International Institute for Environment and Development, accessed 12 September 2024, https://biocultural.iied.org/.

9 G.A. Cajete, *Native Science: Natural Laws of Interdependence*, vol. 315 (Santa Fe, NM: Clear Light Publishers, 2000), 65.

10 Cajete, *Native Science*.

11 Whyte, "On the Role of Traditional Ecological Knowledge as a Collaborative Concept."

12 K.P. Whyte, "What Do Indigenous Knowledges Do for Indigenous Peoples?," in *Traditional Ecological Knowledge: Learning from Indigenous Peoples for Environmental Sustainability*, ed. Melissa K. Nelson and Dan Shilling (Cambridge: Cambridge University Press, 2017), 57–80.

13 Whyte, "What Do Indigenous Knowledges Do for Indigenous Peoples?"

14 Whyte, "On the Role of Traditional Ecological Knowledge as a Collaborative Concept," 5.

15 McGregor, "Traditional Ecological Knowledge and Sustainable Development."

16 Stolomowski quoted in Cajete, *Native Science*, 137.

17 McGregor, "Traditional Ecological Knowledge and Sustainable Development."

18 McGregor, 86.

19 McGregor.

20 D. McGregor, "Traditional Ecological Knowledge and the Two-Row Wampum," *Biodiversity* 3, no. 3 (2002): 8–9.

21 W.A. Wilson, "Introduction: Indigenous Knowledge Recovery Is Indigenous Empowerment," *American Indian Quarterly* 28, no. 3 (2004): 359–72.

22 S.J. Vermeulen, B.M. Campbell, and J.S. Ingram, "Climate Change and Food Systems," *Annual Review of Environment and Resources* 37, no. 1 (2012): 195–222.

23 J. Sbicca, "Growing Food Justice by Planting an Anti-oppression Foundation: Opportunities and Obstacles for a Budding Social Movement," *Agriculture and Human Values* 29, no. 4 (2012): 455–66.

24 E. Holt-Giménez, "Food Security, Food Justice, or Food Sovereignty," in *Cultivating Food Justice: Race, Class, and Sustainability*, ed. A. Hope Alkon and J. Agyeman (Cambridge, MA: MIT Press, 2011), 309–30.

25 I have conflicting thoughts around the terms "food insecurity" and "food security," more specifically with "insecure," as I feel it potentially places a narrative on Indigenous communities that ignores Indigenous sovereignty, and the power dynamics involved in a community becoming "insecure" in the first place. Potential reframed terms could be witnessed under the concept of food "survivance" following Anishinaabe scholar Gerald R. Vizenor, *Survivance: Narratives of Native Presence* (Lincoln: University of Nebraska Press, 2008), who uses "survivance" in opposition to "victimry" narratives that are often placed on Native stories and cultures.

26 D. Evans, D.H. Campbell, and A. Murcott, "A Brief Pre-history of Food Waste and the Social Sciences," *Sociological Review* 60 (2012): 5–26.

27 Evans, Campbell, and Murcott, "A Brief Pre-history of Food Waste and the Social Sciences."

28 One of the biggest learning hurdles has been the proper naming of things. I come from a place of wanting to ensure I name where I come from correctly, and I recognize where I can do this better. For the Huron-Wendat, as "Huron" wasn't a self-given name, it seems "Wendat peoples" would be the most appropriate name; however, "Huron-Wendat" is still commonly used within a lot of the literature gathered.

29 Huron-Wendat Nation, "Brief of the Huron-Wendat Nation: Federal Review of the Environmental and Regulatory Processes Navigation Protection Act," Office of the Nionwentsïo, 7 December 2016, https://www.ourcommons .ca/Content/Committee/421/TRAN/Brief/BR8708093/br-external /HuronWendatFirstNationCouncil-9505010-e.pdf.

30 G.E. Sioui, *Huron-Wendat: The Heritage of the Circle* (Vancouver: UBC Press, 1999).

31 Sioui, *Huron-Wendat*.

32 This is a decolonized methodology comes from Linda Tuhiwai Smith, *Decolonizing Methodologies: Research and Indigenous Peoples*, 3rd ed. (London: Zed Books, 2021), in which names, words, definitions, etc., are placed under a "decolonial lens" and reframed into less harmful terminology. As she states, "Reframing is about taking much greater control over the ways in which Indigenous issues and social problems are discussed and handled. One of the reasons why so many of the social problems which beset Indigenous communities are never solved is that the issues have been framed in a particular way" (175).

33 I want to note that I understand there are models and conversations within waste activism and literature around "closed-loop systems" and "circular economy models." However food-cycling is rooted within the Indigenous knowledge system of Indigenous biocultural heritage. I am not sure if the Indigenous teachings of food cycling can be a part of such systems or if they are to be practised solely in Indigenous food systems; that remains up to the autonomy and sovereignty of the Indigenous communities from which such teachings originate.

34 Cajete, *Native Science*, 72.

35 J. Birch and R.F. Williamson, "Organizational Complexity in Ancestral Wendat Communities," in *From Prehistoric Villages to Cities: Settlement Aggregation and Community Transformation*, ed. J. Birch (New York: Routledge, 2017), 167–92.

36 F.W. Waugh, *Iroquois Foods and Food Preparation* (Ottawa: Government Printing Bureau, 1916).

37 Waugh, *Iroquois Foods and Food Preparation*.

38 Waugh.

39 S.M. Claeson, J.L. Li, J.E. Compton, and P.A. Bisson, "Response of Nutrients, Biofilm, and Benthic Insects to Salmon Carcass Addition," *Canadian Journal of Fisheries and Aquatic Sciences* 63, no. 6 (2006): 1230–41.

40 S.M. Gende, A.E. Miller, and E. Hood, "The Effects of Salmon Carcasses on Soil Nitrogen Pools in a Riparian Forest of Southeastern Alaska," *Canadian Journal of Forest Research* 37, no. 7 (2007): 1194–1202.

41 R. Kirk, *Tradition and Change on the Northwest Coast: The Makah, Nuu-chah-nulth, Southern Kwakiutl and Nuxalk* (Seattle: University of Washington, 1986).

42 P. Nadasdy, "The Gift in the Animal: The Ontology of Hunting and Human-Animal Sociality," *American Ethnologist* 34, no. 1 (2007): 25–43.

43 B. Stevens, "A Recommendation for Polyculture Agriculture to Reduce Nitrogen Loading That Leads to Hypoxia" (MA thesis, University of Maine, 2017).

44 M. Kruse-Peeples, "How to Grow a Three Sisters Garden," Native Seeds Search, 27 March 2016, https://www.nativeseeds.org/blogs/blog-news/how-to-grow-a-three-sisters-garden.

45 A.J. Landon, *The "How" of the Three Sisters: The Origins of Agriculture in Mesoamerica and the Human Niche* (Lincoln: University of Nebraska Press, 2008).

46 T. E. Crews, W. Carton, and L. Olsson, "Is the Future of Agriculture Perennial? Imperatives and Opportunities to Reinvent Agriculture by Shifting from Annual Monocultures to Perennial Polycultures," *Global Sustainability* 1 (2018): e11, doi:10.1017/sus.2018.11.

47 M. Simic, V. Dragicevic, I. Spasojevic, D. Kovacevic, M. Brankov, and Z. Jovanovic, "Effects of Fertilising Systems on Maize Production in Longterm Monoculture," *Fourth International Scientific Symposium Agrosym 2013* (Faculty of Agriculture, University of East Sarajevo), 53–60.

48 A.C. Parker, *Iroquois Uses of Maize and Other Food Plants* (Albany: State University of New York, 1910).

49 Waugh, *Iroquois Foods and Food Preparation*.

50 Waugh, *Iroquois Foods and Food Preparation*.

51 K. Venkat, "The Climate Change and Economic Impacts of Food Waste in the United States," *International Journal on Food System Dynamics* 2, no. 4 (2011): 431–46.

52 K. Schanes, K. Dobernig, and B. Gözet, "Food Waste Matters: A Systematic Review of Household Food Waste Practices and Their Policy Implications," *Journal of Cleaner Production*, no. 182 (2018): 978–91.

53 M. Gooch, A. Felfel, and N. Marenick, "Food Waste in Canada," Value Chain Management Centre, George Morris Centre, November 2010, https://vcm-international.com/wp-content/uploads/2013/04/Food-Waste-in-Canada-112410.pdf; K.M.Parizeau, M. von Massow, and R. Martin, "Household-Level Dynamics of Food Waste Production and Related Beliefs, Attitudes, and Behaviours in Guelph, Ontario," *Waste Management* 35 (2015): 207–17.

54 M.J. Kim and C.M. Hall, "Can Climate Change Awareness Predict Pro-environmental Practices in Restaurants? Comparing High and Low Dining Expenditure," *Sustainability* 11, no. 23 (2019): 6777.

55 R. Fletcher, "Connection with Nature Is an Oxymoron: A Political Ecology of 'Nature-Deficit Disorder,'" *Journal of Environmental Education* 48, no. 4 (2017): 226–33.

56 S.B. Scharper, *For Earth's Sake: Toward a Compassionate Ecology* (Toronto: Novalis, 2013).
57 G.A. Cajete and S.C. Pueblo, "Contemporary Indigenous Education: A Nature-Centered American Indian Philosophy for a 21st Century World," *Futures* 42, no. 10 (2010): 1126–32.
58 Whyte, "On the Role of Traditional Ecological Knowledge as a Collaborative Concept."
59 Nadasdy, "The Gift in the Animal," 25–43.

Chapter Eight

Swimming Upstream against (Neo)colonialism: On Salmon Aquaculture Supremacy and the Decline of Sockeye in the Stó:lō

ERICA HIROKO ISOMURA

Salmon aquaculture operations have been identified as a major cause of the decline of sockeye salmon in the great river known as the Stó:lō, colonially known as the Fraser River. This is recognized by Indigenous communities, environmentalists, and scientists with regards to the spread of disease, pathogens, and sea lice. The state's natural resource management strategies have included a formal Commission of Inquiry into the decline of the Fraser River sockeye (Cohen Commission) and "Aboriginal consultation" regarding salmon aquaculture. However, state-led work in natural resource management fails to recognize the legacies of colonization and the role of Indigenous sovereignty. Drawing upon the critical work of Secwépemc activist and Working Group for Indigenous Food Sovereignty founder Dawn Morrison, this essay identifies the necessity of moving beyond Western scientific methods towards practices of Indigenous governance in order to address the decline in wild salmon. This essay identifies long-term gaps in the environmental sector by exploring the intricate relationships between salmon aquaculture and the colonial-capitalist economy, neoliberal influences on the state, and the reclamation of Indigenous knowledge by and for Indigenous communities.

Introduction

I wrote this essay nearly ten years ago for an undergraduate Indigenous governance course called "State Natural Resource Management and Contemporary Colonialism," taught by Dr. Robyn Heaslip at the University of Victoria. The political environment has changed only slightly since then. Today, instead of a BC Liberal government pushing liquefied natural gas, the Province of British Columbia is led by an NDP government committed to building a dam in the Peace River Valley. Instead of Kinder Morgan's private construction of an oil pipeline, stretching from the Prairies to the Pacific Ocean, that pipeline is now owned by the Government of Canada via the Trans Mountain Corporation, a subsidiary of the Canada Development Investment Corporation.

With consideration of this ongoing, shape-shifting colonialism,[1] I have left the original references to politicians and natural resource extraction projects in this essay as is. In a few more years, the names and faces of these key players may change again, as may the state-supported projects at hand. The natural resource management issues explored below, specifically of the wild salmon decline on the West Coast, cannot be untangled from the matter of historical and ongoing colonialism in British Columbia, and more broadly across North America and elsewhere in the world.

As significant as salmon is to my own diasporic culture, I recognize how the issue of the wild salmon decline is not mine to lead or claim to know what action is best to take; I am after all a non-Indigenous person on these lands. While examining this case study, I have looked to leaders in this field such as Dawn Morrison, co-founder of the Working Group for Indigenous Food Sovereignty and the Wild Salmon Caravan, whom I spoke with during my research. Ultimately, my goal for this essay is to provide research that will aid in considering the interconnectedness of "wild" species, natural resource extraction, globalization, neocolonialism, the state, and settler-colonial environmental politics.

I ascribe much of my learning and guidance to Indigenous academic studies and community. While my undergraduate coursework in environmental studies was primarily rooted in Western scientific methods, I trace much of my political consciousness to anti-racist and decolonial epistemologies shared in Indigenous studies classrooms, and these were further developed among peers and in consultation with written texts by Black, Indigenous, and women of colour authors in student activist spaces.

Today, these lessons continue to influence the work I undertake beyond the walls of the university in my community work, organizing, and writing. Among this work, issues of Indigenous land, space, care and voice still prevail and urgently matter. In *As We Have Always Done*, Michi Saagiig Nishnaabeg scholar Leanne Betasamosake Simpson writes, "This is our struggle. It is hard work. It is responsibility and accountability and real, and you cannot get a degree in it. It's not a hashtag or a sticker or a t-shirt or a selfie or anything white liberals think is important. It isn't recognition. It is struggle." [2]

With the contribution of this essay, I hope to shift perspectives so as to envision a better way forward and more fully understand the complex history of the environmental issues we face today – in this case, the decline of wild salmon. To me, there is an urgent need for the environmental movement to firmly ground its commitment to Indigenous liberation and self-determination as ends in themselves, not just as means to combat climate catastrophe. Far more work on these matters has been written by Indigenous scholars, thinkers, and writers that speak to these critical issues of governance. For all of us, as is true of so many other interrelated matters of climate and social justice of our time, this work will take a crucial rethinking and re-structuring of what it means to be in relationship with land, waterways, and each other. There is so much more to be said and done, but for now I begin here.

Case Study: Industrial Aquaculture and the Decline of Wild Salmon in the Stó:lō

The salmon are the bloodline of our people, our main food and without them we lose our rights to fish! When they tested our ancestors' bones they showed that salmon made up 95% of our diet! With the salmon decreasing in our diet, we are suffering from disease, like diabetes and cancer. Our lives are interconnected with the salmon, they are our relatives. The salmon are going through the same colonial experience that we have gone through.[3]

As declared by the Indigenous Salmon Defenders,[4] wild salmon are one of the most important foods across what is now known as British Columbia.[5]

The decline of traditional foods as a result of the impacts of colonization, including development, urbanization, and growth of the market economy, has contributed to increasing food insecurity and food-related illnesses within Indigenous communities.[6] The decline of traditional foods, including wild salmon stocks, has been widely recognized as problematic. In 2009, the Canadian governor general in council affirmed this issue by establishing the Cohen Commission of Inquiry.[7]

Many salmon aquaculture operations are located on the northwestern coast of British Columbia along ocean migration routes where Pacific salmon pass, headed for the Stó:lō, colonially known as the Fraser River.[8] Stó:lō translates to "river," "river of rivers," and "tribe of tribes" in the Halq'eméylem language,[9] which is spoken by caretakers of the land upon which I was born and raised. This is the longest river in the province and has historically been considered the greatest salmon-producing river in the world.[10] Ten years ago, when the Cohen Commission began, more than a million sockeye salmon travelled home along this river, making the long journey from the Pacific Ocean to Secwépmec territories in the BC interior and other spawning grounds along the way. In 2020, the Department of Fisheries and Oceans (DFO) announced that the return of sockeye to the Stó:lō would result in an estimated 283,000 salmon, one of the lowest returns of sockeye salmon in history.[11] This news led First Nations leaders to officially "declare collapse of pacific sockeye."[12]

Indigenous communities, environmentalists, and scientists have long recognized the salmon aquaculture industry as one of the causes of the decline of wild salmon due to its spread of disease, pathogens, and sea lice.[13] Through my research on this topic, I was most affected by the voices of Dawn Morrison, who chairs the Working Group on Indigenous Food Sovereignty and organizes the Wild Salmon Caravan (previously known as the Indigenous Salmon Defenders March for Wild Salmon), and Dr. Alexandra Morton, a marine biologist. While both of these women have been actively involved with wild salmon activism, they approach the issue from two different perspectives. Morton is a white settler who moved to British Columbia from the United States in the 1980s. She has endeavoured to use Western science in order to prove that diseases incurred by the industry are sufficient reason for the DFO to shut down the salmon farms.[14] By contrast, Morrison, as a Secwépemc woman, counts salmon among her relatives. She believes that Indigenous governance plays a stronger duty in this matter than science,[15] which challenged me to make deeper connections between this case study, colonial histories, and the role of Indigenous governance.

Resource Management and Colonial Legacies

Wild Pacific salmon are regulated as a resource by the DFO. The politics of resource management in Canada are complex – this is perhaps an understatement, considering the largely unacknowledged occupation of Indigenous ancestral homelands by the state and its occupants. This includes many settlers, even those who have identified capitalism and privatization of lands as corrupt.[16] In Canada, natural resource management excludes Indigenous governance for a multitude of reasons, including settler Canadians' unwillingness to accept its precedence and instead favouring colonial control and world views.[17] Despite co-management strategies in land and resource management efforts as an attempt to "empower" Indigenous communities, there is a lack of dialogue in the scientific realm on power dynamics, politics, ethics, and history.[18] These efforts to consult may become what Nadasdy describes as a "subtle extension of empire" since they do not fully embrace Indigenous knowledge as an entity in and of itself and fail to understand complex Indigenous relationships with the land, including animals such as salmon.[19]

Colonizers and settlers did not (and still do not) easily comprehend Indigenous world views and how they are deeply embedded in cultural, social, political, and economic values.[20] A clear example of this can be seen in the historical treaty-making processes across Canada, including the agreement with Governor James Douglas on Vancouver Island, who did not understand the local peoples' stewardship of the land since it did not appear to be "occupied by cultivation, or had houses build on."[21] To Douglas, all other land was "to be regarded as waste, and applicable to the purposes of colonization."[22] Many of these misconceptions continue to prevail, and therefore, amid discussions of environmental concern, such as the salmon decline, Indigenous people are reduced to mere stakeholders, alongside government agencies, scientists, environmentalists, and fishers, instead of acknowledged and affirmed as knowledge holders and caretakers whose lives have been interwoven with the salmon since time immemorial.[23]

The diminished status of Indigenous fishing is a colonial legacy of British Columbia.[24] Colonizers' ignorance of the relationships between Indigenous people and salmon, in addition to the colonial assumptions regarding Western superiority, have fostered a power imbalance since the initiation of the first commercial salmon fisheries in the province. The first salmon canneries emerged in coastal British Columbia in the 1870s, long after Indigenous peoples had established salmon-based economies.[25] As the fishing industry expanded, work opportunities arose for Indigenous people at canneries, which enabled a mixed economy, serving as a means to support concurrent Indigenous fisheries and traditional economies.[26] The fishing industry was not pleased with fishing competition from Indigenous people, and the state acknowledged the successes of this new mixed economy with disdain.[27] This led to the 1888 Canadian fisheries regulations prohibiting Indigenous net and spear fishing practices without government licences, while simultaneously permitting the "liberty" of Indigenous peoples to fish for sustenance.[28] One could consider this a form of systemic gaslighting. These laws increased in the 1890s and through the 1900s, when Indigenous net fishing practices were blamed for salmon run declines and conservation efforts were instated.[29] The limitations on Indigenous fishing practices subsequently provided increasing resources to settlers.[30] Colonizers consistently undermined Indigenous economies and ways of knowing for their own benefit, which is an issue that still prevails today.

The impact of colonization on Indigenous fishing rights is unsurprising, given Canada's colonial legacy and ongoing relationships with

Indigenous peoples, communities, and nations. While this example demonstrates an obvious historical advantage given to settlers, it is crucial to recognize that colonial impositions and oppressive power relations have not disappeared – they are simply less blatant today.[31] As described by Taiaiake Alfred and Jeff Corntassel, "we live in an era of postmodern imperialism and manipulations by shape-shifting colonial powers; the instruments of domination are evolving and inventing new methods to erase Indigenous histories and senses of place."[32]

Shape-shifting colonialism enables the loss of culture through Western hegemonies that normalize injustices for Indigenous people,[33] such as the dispossession of Indigenous homelands for the extraction of natural resources.[34] As stated by Simpson, "we have not had the right to say no to development because Indigenous communities are not seen as people. They are seen as resources."[35] The present-day colonial reality is compounded by countless injustices and wrongdoings in the name of greed and empire.

Political Strategy, Multinational Corporations, and Farmed Fish

The decline of wild salmon in the Stó:lō and its relevance to salmon aquaculture provides a crucial example of the role that colonial powers play in diminishing Indigenous economies and histories in the contemporary context. As argued by Newell,[36] Indigenous control over fisheries were and remain crucial to the well-being of Indigenous economies and the self-sufficiency of Indigenous communities. However, this is undermined by the impacts of salmon aquaculture in British Columbia.

Today, salmon aquaculture plays a large role in the BC economy. Farmed salmon is its largest agricultural export product,[37] and ecological externalities excluded, it is significantly more profitable than the wild salmon industry.[38] In 2007, the salmon farming industry contributed more than double the value of the wild salmon industry to British Columbia's GDP: $134 million compared to $67 million, respectively.[39] For the salmon aquaculture industry and the Province of British Columbia, this has been a great success.[40]

This economic achievement has been a result of many actors on both the local and global levels. One of these actors was former Prime Minister Brian Mulroney of the Progressive Conservatives, who replaced the *Foreign Investment Review Act* with the *Investment Canada Act* in 1985, which withdrew the requirement that Canadian citizens must hold majority ownership of Canadian-registered companies.[41] Mulroney's decision was a political strategy to encourage Canada's stake in the globalized economy,[42] and was rooted in neoliberal ideologies, which encourage self-regulated market growth and expansion amid the absence of state interference and regulation. Neoliberalism further allows production and distribution to materialize on a global scale.[43] This process of globalization may be an example of the aforementioned shape-shifting colonialism as "a deepening, hastening and stretching of an already-existing empire,"[44] or neocolonialism, as coined by Nkrumah,[45] where multinational corporations reap profits from dispossessed Indigenous homelands.

As pointed out by Dr. Isabel Altamirano-Jiménez, neoliberalism and neocolonialism are "not driven by an external, invisible hand but by specific actors, sites, institutions, networks, the state, and discourses, all of which have material effects in different places."[46] Mulroney's decision is noted here because it enabled the establishment of Norwegian salmon aquaculture companies in Canada, which controlled over 90 per cent of annual farmed salmon production in this country as of 2006.[47] While Mulroney was instating the new investment act, there was an outbreak of parasites in Norwegian salmon aquaculture, which led to the Norwegian government's enforcement of stricter rules for aquaculture operations.[48]

These concurrent policy changes in both Canada and Norway provided an opportunity for Norwegian companies to move their operations and practices to Canada, specifically coastal British Columbia, which would provide an "exemplary physical and biological habitat"[49] for salmon farming. The year prior, the DFO began to allow fisheries to import Atlantic salmon eggs, despite widespread concern among the public regarding the introduction of non-native species.[50] This also served the goals of Norwegian companies, which had already established an international export market for Atlantic salmon.[51] While Norwegian companies became established in British Columbia, the province's salmon farming industry expanded its production of Atlantic salmon, which made up 76 per cent of British Columbia's farmed aquaculture by 2006. Both scientists and Indigenous community members believe the aquaculture industry increased the transmission of sea lice, disease, and pathogens from Atlantic salmon to wild Pacific salmon due to the close proximity of the farms to wild salmon runs.[52] The implications of these impacts, which profit foreign multinational corporations, suggest an embedded neocolonialism, as these corporations "reap the benefit of similar colonial endeavors throughout the world."[53]

The Many Interchangeable Bodies and Faces of (Neo)colonialism

British Columbia is now dependent on the economic value that industrial aquaculture provides – the influence of neoliberalism on the state has become intertwined with its own economic drive and duty to support a population.[54] This was embodied by the BC Liberals' slogan, "Secure Economy, Secure Tomorrow,"[55] with party leader Christy Clark remaining adamant about the necessity of economic growth.[56] News media during Clark's leadership suggested that she was in favour of Kinder Morgan, the Enbridge Northern Gateway Pipeline, and various liquified natural gas projects.[57] She promised to provide "benefits" to First Nations communities,[58] yet many Indigenous nations did not want to put land or water at risk for these developments.[59] Clark's words echoed those of the DFO in its policy principles, which state that the federal government will "respect constitutionally protected Aboriginal and treaty rights and will work with interested and affected Aboriginal communities to facilitate their participation in aquaculture development."[60] These examples exhibit how the state's inclusion of Aboriginal rights and title in natural resource development is based on a narrow belief that Indigenous people desire to be part of the globalized market economy through what Altamirano-Jiménez refers to as a configuration of a new market-based Indigenous citizenship.[61]

This case study shows that many Indigenous people disproportionately benefit from these economic developments and remain systemically oppressed by globalization.[62] While many Indigenous communities on northern and mid-Vancouver Island have participated in salmon aquaculture projects, there remains contention within communities about the viability and sustainability of these projects, in addition to the way they continuously fail to meet these communities' cultural and spiritual needs. Ironically, the rise of aquaculture was intended to meet the goals of the "blue revolution" – namely, to provide protein to more people.[63] Evidently, the nature of shape-shifting colonizers in power (e.g., politicians, government agencies, corporations, band council governments) shows that these actors may wear the masks of many different interchangeable bodies. As articulated by Khelsilem Rivers, these actors are trivial and are not themselves the enemy; rather, they are the symptoms of the enemy, which is ultimately colonialism.[64]

Indigenous Food Sovereignty and an Anti-racist Environmental Movement

The work of Indigenous resurgence, cultural reclamation, anti-racism, and oppression often appears very different from environmental activism on, for example, such issues as conservation or biodiversity loss. Yet it is these siloed solutions and ways of thinking that are holding us back from making lasting environmental change. Indigenous activists such as Morrison lead the way towards multifaceted alternatives to Western scientific approaches to engage and educate the public, policymakers, researchers, and regional governments.[65] Beyond her focus on salmon as a food source and her concern for the species' ecological health, Morrison's work is spiritually and creatively nuanced – her leadership in organizing the Wild Salmon Caravan, a celebratory parade and ceremony to call wild salmon home, has brought together a vast network of Indigenous people, settlers, environmentalists, food activists, artists, organizations, and communities, both urban and rural, along the river.[66]

Further, through Morrison's work with the Working Group on Indigenous Food Sovereignty, she challenges extractivism,[67] using anti-colonial strategies to reclaim Indigenous knowledge.[68] Working alongside a network led by urban Indigenous women, including Cease Wyss, Anne Riley, and Jolene Andrew, Morrison is actively recovering, practising, and maintaining Indigenous food systems in a contemporary context.[69] Morrison's work embodies a resurgent Indigenous movement through a decolonized diet, in which "the struggle for freedom is the reconstitution of our own sick and weakened physical bodies and community relationships accomplished through a return to the natural sources of food … lived by our ancestors."[70]

Vanessa Watts writes, "The border where human-as-the-centre begins still exists and continues to determine the bounds for capacity and action."[71] From an environmental perspective, I would add that the border where *white*-as-the-centre begins limits the imagination and potential for capacity and action in the sphere of climate justice, whether in regards to issues of salmon, pipelines, or otherwise. Today, a number of grassroots Indigenous organizations and people work to educate others and support Indigenous claims to traditional practices and diets.[72] There is much potential for supportive, respectful, and collaborative contributions from settlers within this work, whether through financial reparations or through volunteer labour and support.

Colonialism is not just an Indigenous issue. Intertwined with oppressive racism, gendered violence, poverty, environmental justice, and war,[73] colonialism's violent consequences "enclos[e] whiteness into the centre."[74] While it may not be immediately apparent how these social issues relate to a decline in salmon stocks in a river, all of these social and environmental issues, including the decline of wild salmon, are rooted in the structures and narratives of a settler-colonial society: what we value, how we behave, and our ways of thinking and relating to one another. Settler colonialism, like racism, is a world view that we are raised in – it's in the air we breathe.[75] Beyond simply recognizing past colonial impacts and oppressions, there is both an opportunity and a need to centre these conversations in environmental work and to follow the leadership of Indigenous communities. The process of undoing colonial harm will be a long, hard journey. It will take much candour and creativity; it will require risk taking and an embrace of the unknown. We may not see it through to the very end in our own lifetimes, but the salmon are counting on us. Every year, no matter how many, the salmon are still swimming upstream.

Acknowledgments

Thank you to Dawn Morrison, the Working Group on Indigenous Food Sovereignty, and the Wild Salmon Caravan for leading this work to defend wild salmon. I am grateful for the friendship and mutual support extended by my peers in the student-run groups the Students of Colour Collective and the Native Students Union, which provided me with a means of enduring white supremacy during my time in academia and offered space to contribute my voice. Thank you to professors Dr. Christine O'Bonsawin, Dr. Rita Dhamoon, Dr. Robyn Heaslip, Dr. Cliff Atleo Jr., and Dr. Jessica Dempsey, all of whom helped challenge, guide, and support my work during my undergraduate years at the University of Victoria.

NOTES

1 T. Alfred and J. Corntassel, "Being Indigenous: Resurgences against Contemporary Colonialism," *Government and Opposition: Politics of Identity* 9 (2005): 597–614.
2 L. Simpson, *As We Have Always Done: Indigenous Freedom through Radical Resistance* (Minneapolis: University of Minnesota Press, 2020), 67.
3 Indigenous Salmon Defenders, "Call to Indigenous Salmon Defenders! Virus Threat to Indigenous/Wild Salmon! Warning for Indigenous Peoples and Fishers!," *Native News North* [E-mail listserv], accessed 12 December 2013, https://groups.yahoo.com/neo/groups/NatNews-north /conversations/topics/22556.
4 Indigenous Salmon Defenders, "Call to Indigenous Salmon Defenders!"
5 D. Morrison, "Indigenous Salmon Defenders – Media Release," 28 February 2013, accessed 13 December 2013, http://groups.yahoo.com/neo/groups /NatNews-north/conversations/topics/24373; "Traditional Food Fact Sheets: Nutrition," First Nations Health Council, accessed 22 October 2024, https://www.fnha.ca/Documents/Traditional_Food_Fact_Sheets.pdf
6 Morrison, "Indigenous Salmon Defenders"; N.J. Turner and K.L. Turner, "Traditional Food Systems, Erosion and Renewal in Northwestern North

America," *Indian Journal of Traditional Knowledge* 6, no. 1 (2006): 57–68; Indigenous Salmon Defenders, "Call to Indigenous Salmon Defenders!"

7 Cohen Commission of Inquiry into the Decline of the Sockeye Salmon in the Fraser River, "Chapter 8: Salmon Farm Management," in *Final Report*, vol. 1, *The Sockeye Fishery* (Ottawa: Privy Council, 2012). Henceforth cited as Cohen Commission, "Salman Farm Management."

8 J. Volpe and K. Shaw, "Fish Farms and Neoliberalism: Salmon Aquaculture in British Columbia," in *Environmental Challenges and Opportunities: Local-Global Perspectives on Canadian Issues*, ed. C. Gore and P. Stoett (Toronto: Edmond Montgomery, 2012), 131–58; A. Morton in *Salmon Confidential*, dir. Twyla Toscovich (Sointula, BC: Pacific Coast Wild Salmon Society, 2013).

9 Stó:lō Nation, "Stó:lō: A Note on the Halq'eméylem Language," Stó:lō Research and Resource Management Centre, accessed 19 September 2024, http://www.srrmcentre.com/StoneTxwelatse/06NoteHalq.html.

10 R.L. Burgner, "Life History of Sockeye Salmon (Oncorhyhchus Nerka)," in *Pacific Salmon Life Histories*, ed. C. Groot and L. Margolis (Vancouver: UBC Press, 1991), 1–117.

11 Fisheries and Oceans Canada, *Aquaculture in Canada 2012: A Report on Aquaculture Sustainability* (Ottawa: Fisheries and Oceans Canada, 2012).

12 "First Nations Leaders Declare Collapse of Pacific Sockeye," Union of BC Indian Chiefs, 18 August 2020, https://www.ubcic.bc.ca/first_nations_leaders_declare_collapse_of_pacific_sockeye.

13 Cohen Commission, "Salmon Farm Management"; Indigenous Salmon Defenders, "Call to Indigenous Salmon Defenders!"; Morton, *Salmon Confidential*; Volpe and Shaw, "Fish Farms and Neoliberalism."

14 Morrison, "Indigenous Salmon Defenders."

15 D. Morrison, personal communication, 13 November 2013.

16 E. Tuck and W. Yang, "Decolonization Is Not a Metaphor," *Decolonization: Indigeneity, Education & Society* 1, no. 1 (2012), https://jps.library.utoronto.ca/index.php/des/article/view/18630/15554.

17 L. Simpson, "Anti-colonial Strategies for the Recovery and Maintenance of Indigenous Knowledge," *American Indian Quarterly* 28, nos. 3–4 (2004): 373–84.

18 P. Nadasdy, "The Anti-politics of TEK: The Institutionalization of Co-management Discourse and Practice," *Anthropologica* 47 (2005): 215–32; Simpson, "Anti-colonial Strategies."

19 Nadasdy, "The Anti-politics of TEK," 228.

20 First Nations Studies Program, "Aboriginal Fisheries in British Columbia," Indigenous Foundations, University of British Columbia, accessed 20 September 2024, https://indigenousfoundations.arts.ubc.ca/aboriginal_fisheries_in_british_columbia/ ; R. Kuohhanen, "Indigenous Economies, Theories of Subsistence, and Women: Exploring the Social Economy

Model for Indigenous Governance," *American Indian Quarterly* 35, no. 2 (2011): 212–40; L. Little Bear, "Jagged Worldviews Colliding," in *Reclaiming Indigenous Voice and Vision*, ed. M. Battiste (Vancouver: UBC Press, 2011), 77–85.

21 P. Tennant, "The Douglas Treaties and Aboriginal Title," in *Aboriginal Peoples and Politics: The Indian Land Question in British Columbia, 1849–1989* (Vancouver: UBC Press, 1990), 18.

22 Tennant, "The Douglas Treaties and Aboriginal Title," 18.

23 S. Hsu, "Salmon Farming in British Columbia," University of British Columbia Faculty of Law, accessed 13 December 2013, http://www.law.ubc .ca/files/pdf/enlaw/salmonfarming04_20_09.pdf; First Nations Studies Program, "Aboriginal Fisheries in British Columbia."

24 D.C. Harris, *Fish, Law, and Colonialism: The Legal Capture of Salmon in British Columbia* (Toronto: University of Toronto Press, 2001); D. Newell, *Tangled Webs of History: Indians and the Law in Canada's Pacific Coast Fisheries* (Toronto: University of Toronto Press, 1993).

25 Newell, *Tangled Webs of History*; First Nations Studies Program, "Aboriginal Fisheries in British Columbia."

26 Newell, *Tangled Webs of History*; First Nations Studies Program, "Aboriginal Fisheries in British Columbia"; Kuohhanen, "Indigenous Economies, Theories of Subsistence, and Women."

27 First Nations Studies Program, "Aboriginal Fisheries in British Columbia."

28 First Nations Studies Program, "Aboriginal Fisheries in British Columbia"; Newell, *Tangled Webs of History*.

29 First Nations Studies Program, "Aboriginal Fisheries in British Columbia."

30 First Nations Studies Program.

31 S. Irlbacher-Fox, *Finding Dashaa: Self Government, Social Suffering, and Aboriginal Policy in Canada* (Vancouver: UBC Press, 2009).

32 Alfred and Corntassel, "Being Indigenous," 601.

33 Irlbacher-Fox, *Finding Dashaa.*

34 N. Klein "Dancing the World into Being: A Conversation with Idle No More's Leanne Simpson," *YES! Magazine*, 6 March 2013, https://www .yesmagazine.org/social-justice/2013/03/06/dancing-the-world-into-being -a-conversation-with-idle-no-more-leanne-simpson.

35 Klein "Dancing the World into Being."

36 Newell, *Tangled Webs of History.*

37 "Farmed Salmon," Fisheries and Oceans Canada, last modified 15 March 2017, https://www.dfo-mpo.gc.ca/aquaculture/sector-secteur/species -especes/salmon-saumon-eng.htm; Columbia Department of Agriculture and Seafood, *2019 International Export Highlights Year in Review* (Victoria: British Columbia Department of Agriculture and Seafood, 2019), https:// www2.gov.bc.ca/assets/gov/farming-natural-resources-and-industry

/agriculture-and-seafood/statistics/market-analysis-and-trade-statistics
/2019_bc_agrifood_and_seafood_export_highlights.pdf.

38 Hsu, "Salmon Farming in British Columbia"; U.R. Sumaila, *Aquaculture Economics: The B.C. Experience* (Vancouver: UBC Fisheries Centre, 2005), http://www.whoi.edu/fileserver.do?id=6801&pt=2&p=9550.

39 Legislative Assembly of British Columbia, Special Committee on Sustainable Aquaculture, *Final Report*, vol. 2, *Economic Impacts and Prospects of the Salmon Farming and Wild Salmon Industries in British Columbia* (Victoria: Legislative Assembly of British Columbia, 2007, https://www.wildbcsalmon.ca/Down /files/AQUACULTURE-38-3-Vol2.pdf.

40 Fisheries and Oceans Canada, *Aquaculture in Canada 2012*.

41 Volpe and Shaw, "Fish Farms and Neoliberalism."

42 Canadian Press, "Brian Mulroney: Canada Needs Foreign Investment," *CTV News*, 24 October 2012, https://www.ctvnews.ca/canada /brian-mulroney-canada-needs-foreign-investment-1.1008276.

43 J. Ervin and Z.A. Smith, *Globalization: A Reference Handbook* (Santa Barbara, CA: ABC-CLIO, 2008).

44 Alfred and Corntassel, "Being Indigenous," 601.

45 K. Nkrumah, *Neo-colonialism: The Last Stage of Imperialism* (New York: International Publishers, 1966).

46 I. Altamirano-Jiménez, "Indigeneity, Nature, and Neoliberalism," in *Indigenous Encounters with Neoliberalism: Place, Women and the Environment in Canada and Mexico* (Vancouver: UBC Press, 2013), 70.

47 Volpe and Shaw, "Fish Farms and Neoliberalism."

48 Volpe and Shaw.

49 Volpe and Shaw, 5.

50 P.A. Robson, *How We Got Here: A Brief History of Salmon Farming in B.C. Salmon Farming: The Whole Story* (Nanoose Bay, BC: Heritage House Publishing, 2006).

51 Volpe and Shaw, "Fish Farms and Neoliberalism."

52 Volpe and Shaw, "Fish Farms and Neoliberalism"; Morton, *Salmon Confidential.*

53 G.A. Slowey, "Globalization and Self-Government: Impacts and Implications for First Nations in Canada," *American Review of Canadian Studies* 31, nos. 1–2 (2001): 265–81.

54 I. Altamirano-Jiménez, "North American First Peoples: Slipping Up into Market Citizenship?," *Citizenship Studies* 8, no. 4 (2004): 349–65.

55 J. Fowlie and L. Culbert, "B.C. Liberal Platform Promises Debt Reduction but Offers Few New Ideas," *Vancouver Sun*, 14 April, 2013, https://vancouversun.com/news/bc%20election /bc-liberal-platform-promises-debt-reduction-but-offers-few-new-ideas.

56 J. Hunter, "Economic Factors Mean B.C. Government Unlikely to Oppose
   Kinder Morgan Bid," *Globe and Mail*, 15 December 2013,
   http://www.theglobeandmail.com/news/british-columbia/economic
   -factors-mean-bc-government-unlikely-to-oppose-kinder-morgan-bid
   /article15978528/.
57 Hunter, "Economic Factors"; I. Baily and B. Jang, "Clark, Redford
   Have Framework for Pipeline Deal," *Globe and Mail*, 5 November 2013,
   http://www.theglobeandmail.com/news/national
   /clark-redford-reach-deal-on-pipelines/article15260483/.
58 M. Wright, "B.C. and Alberta Now Agree: These Five Conditions
   Must Be Met for Pipelines to Go Ahead," *Globe and Mail*, 5 November 2013,
   https://www.theglobeandmail.com/news/british-columbia/bc-and
   -alberta-now-agree-these-conditions-must-be-met-for-pipelines-to-go-ahead
   /article15273829/.
59 N. Peterson, "Save the Fraser Declaration – Alberta and Northwest
   Territories First Nations Join BC First Nations in Opposing Enbridge
   Pipeline and Tankers Proposal," West Coast Environmental Law, 31 January
   2012, https://www.wcel.org/blog/save-fraser-declaration-alberta-and
   -northwest-territories-first-nations-join-bc-first-nations.
60 Cohen Commission, "Salmon Farm Management," 392.
61 Kuohhanen, "Indigenous Economies, Theories of Subsistence, and
   Women"; Altamirano-Jiménez, "North American First Peoples."
62 Alfred and Corntassel, "Being Indigenous.
63 Volpe and Shaw, "Fish Farms and Neoliberalism."
64 K. Rivers, "Enemies: Left and Right," *DividedNoMore*, 11 November 2013,
   http://dividednomore.ca/2013/11/11/enemies-left-and-right/.
65 D. Morrison, personal communication, 13 November 2013, 10 September
   2019, and 7 February 2020.
66 I. Marcuse, "Wild Salmon Caravan Vancouver Parade – Murray
   Bush Photos," *Grandview Woodland Food Connection*, 9 October 2017, https://
   gwfoodconnection.com/2017/10/09/wild-salmon-caravan-vancouver
   -parade-murray-bush-photos/; E. Isomura, "Photos: Wild Salmon Caravan
   2018," *Vancouver Neighbourhood Food Networks*, October 2018, http://
   vancouverfoodnetworks.com/2018/10/wild-salmon-caravan-2018/.
67 Klein "Dancing the World into Being."
68 Simpson, "Anti-colonial Strategies."
69 D. Morrison, personal communication, 7 February 2020.
70 Alfred and Corntassel, "Being Indigenous," 613.
71 V. Watts, "Indigenous Place-Thought & Agency amongst Humans and Non-
   humans (First Woman and Sky Woman Go On a European World Tour!),"
   Decolonization: Indigeneity, Education & Society 2, no. 1 (2013): 20–34.

72 "Indigenous Food Systems Network," Working Group on Indigenous Food Sovereignty, accessed 20 September 2024, http://www
.indigenousfoodsystems.org/.

73 H. Walia, "Decolonizing Together: Moving beyond a Politics of Solidarity to a Practice of Decolonization," in *Organize! Building from the Local for Global Justice*, ed. A. Choudry, J. Hanley, and E. Shragge (Oakland, CA: PM Press, 2012).

74 Watts, "Indigenous Place-Thought."

75 A. Velshi, "Alicia Garza: Racism Is 'Everywhere, It's Like the Air We Breathe,'" *MSNBC*, 2 January 2021, https://www.msnbc.com/ali-velshi
/watch/alicia-garza-racism-is-everywhere-it-s-like-the-air-we
-breathe-98700357885.

PART THREE

STRUGGLES

Thieves of the North-West Coast: Understanding Indigenous and Non-Indigenous Relations in Clayoquot Sound, 1791–1792

ANDRÉ BESSETTE

Intent

In many ways, Canadian history has continued the colonization of Turtle Island, whether through the concealment of genocidal practices by state, church, and residential schools, the erasure of Indigenous land and food-management systems, and by revising or ignoring the negative impacts that Western colonization has perpetrated on Turtle Island and across the globe.

I wrote this essay several years ago. I have grown, I have been gifted knowledge, and I have gained experience since then. Looking back, I was influenced by the inherent racism in the academic system and my upbringing in Canadian society. I've witnessed systemic racism and experienced anti-Indigenous behaviour in professional and personal settings. Yet, I do not want to revise my history by destroying this record of growth and change. Where I have made a change to the original essay, I have underlined my original words (and provided new ones in brackets). In this way, I attempt to continue to reject the Western "tendency to compartmentalize experience" that Umeek (Richard Atleo) speaks about in *Tsawalk: A Nuu-chah-nulth Worldview*.[1]

Thieves of the North-West Coast is an attempt to Indigenize the undergraduate history essay format and decolonize deductive reasoning. I attempted to bridge the historical primary documents written by non-Indigenous Americans in the early 1700s with modern Indigenous philosophy written by a Nuu-chah-nulth hereditary chief. Umeek's (Richard Atleo's) book *Tsawalk: A Nuu-chah-nulth Worldview* shares the culture, language, and world view of the Nuu-chah-nulth in both Indigenous and Western ways.[2]

Figure 9.1. André Bessette performs the Red River Jig at the annual Louis Riel Day celebration with Compaigni V'ni Dansi.

Source: Courtesy of Chris Randle. Photo property of Compaigni V'ni Dansi.

I tried to respect and recognize *Tsawalk* by grounding myself in my Métis culture, connecting with Indigenous colleagues and sources, and refusing to filter these stories into a single, objective thematic statement. Considering the restrictions of an undergraduate degree, I considered this to be the best way that I could follow protocol while writing about the Nuu-chah-nulth Nation's unceded stories as they were written about by Americans during the invasion.

At the very least, I hope this essay points to the vitality of the Nuu-chah-nulth world view in their language and people today as it was in the

1700s.³ Sadly, the threat of revisionism, kleptomania, and deadly force present in Western world views is still as potent as ever.

Introduction

The early histories of interactions on the North-West Coast abound with stories of thieving "savages," failed settlements, and conflicts over trade routes that could have ensured advantages for both settler and Indigenous interests. However, primary documents used to recount these histories, such as *Voyages of the Columbia to the Northwest Coast 1787–1790 and 1790–1793*, are an incomplete telling of the story (full of racism and anti-Indigenous sentiment).⁴

For this essay I desired to develop a more distinct understanding of the interactions of theft between the Americans on the *Columbia* and the Tla-o-qui-aht peoples on what is now known as Vancouver Island. Robert Haswell, the first mate of the *Columbia*, a ship that sailed to the North-West Coast twice, responded to his first encounter with Indigenous people of the area with "[they are] a smart sett if active fellows but like all others without one exception on this Coast addicted to thefts."⁵ This was the first time that Haswell had ever been to the North-West Coast, but he had a preconceived notion that the Indigenous peoples of this area were predisposed to thievery. While I do not believe Haswell's negative (racist) generalization about North-West Coast peoples, I also question the assumption that all sea-going traders "had no interest in cultivating the good will of the natives."⁶ These early narratives depict a significant amount of theft committed by both settler and Indigenous peoples, but in the end, the theft of land and possessions ultimately benefited the non-Indigenous settlers (and continues to benefit generations of settlers without reparations for stolen sovereign territories).

These relations of mutual thievery can be used to further highlight the complexities of two very different societies interacting for one of the first times. The circumstances were sometimes deadly, driven by socioeconomic factors, and confused by cultural differences and languages barriers. At times, sailors took food from the surrounding area without understanding that they were stealing from local resources. At others, First Nations were recorded as blatantly taking European objects as if they were their own. The intentions behind and reactions to these thefts revealed that each group followed different systems of ownership.

Self-Introduction

As a Métis and mixed settler person I have attempted to Indigenize this process of history with the means there are available to me. This essay represents the final project that a history major must complete to achieve their bachelor of arts. My restrictions were to focus on a primary document from the period, then present a fifteen-page article within one semester while juggling multiple courses. It is not as if my Métis ancestry simply allows me to write about the Tla-o-qui-aht First Nation without complication or appropriation. Since I had no prior connection to the Tla-o-qui-aht First Nation or Nuu-chah-nulth Nation, on top of the circumstances above, meant that I would have breached protocol, in my opinion, if I approached either with this project. I do not feel that I was in a suitable position to respectfully record the Nuu-chah-nulth's oral histories and traditions (and this project offered no clear benefit to the nation).

I was left to obtain these histories from the restrictive realm of textual documents, which is dominated by Eurocentric beliefs and processes. During this research I did not come across a single historian who used the oral traditions of the Nuu-chah-nulth to identify, analyse, or clarify early contact histories in Clayoquot Sound or surrounding regions. Even more contemporary authors, such as Daniel Clayton, fail to effectively re-establish Indigenous presence, authority, and voice in their histories of the North-West Coast.[7] This is because Daniel Clayton is a British man writing a part of the Nuu-chah-nulth people's history without directly involving them in the process.

Integrating Indigenous oral history into academia will continue to be next to impossible if certain assumptions are upheld by academics and settler society. The assumption that I am most concerned with is that the multitude of nations, languages, and cultures stumbled upon by Europeans in what is now known as the Americas have disappeared since contact. This is untrue. Cultures grow and change. Their histories, world views, and ideologies have not been lost. Cultural values are upheld, strengthened, and in some cases adapted for new generations using oral history. Yet, despite this strength and resilience, I have been unable to locate works of academic histories of First Nations that actively integrate oral histories. To do so would be to re-envision the past without relying on Eurocentric documentation. This would have an Indigenizing effect on history, by which I mean it would facilitate the integration of Indigenous perspectives, histories, and ideologies into mainstream accounts of the history of the North-West Coast.

Tsawalk: A Nuu-chah-nulth World View

I was in the same position as other historians until I became aware of Umeek's (Dr. Richard Atleo's) 2005 book *Tsawalk*.[8] Umeek is a Nuu-chah-nulth hereditary chief and the first Indigenous person to earn a doctoral degree in British Columbia, and he carries his culture's oral histories and traditions while simultaneously integrating their world view of *Tsawalk* into academic philosophy. To honour Nuu-chah-nulth culture and oral histories and to equalize the Eurocentric documentation used in this essay, I will incorporate *Tsawalk* to inform my analytical approach.

Tsawalk is not only an ideology; it is also an analytical methodology. Umeek refers to the concept of *Tsawalk* as "a unity, or meaningful inter-relationship, between all variables of existence."[9] By accepting the world as irreversibly interrelated one must also accept that a single explanation for anything is impossible. Since *Tsawalk* respects both physical and meta-physical realms of reality, it recognizes the relationship between physical and spiritual bodies. Umeek suggests that while the metaphysical part of life may be difficult to describe, it can be acknowledged through practice and ceremony.[10]

Related to the practice of *Tsawalk* is *isaak*: the respect for all living things "predicated upon the notion that every life form has intrinsic value and that this should be recognized through appropriate protocols of interaction."[11] Umeek explains that the wolf shows respect to the deer by not tearing out its innards after a kill; he states that the Nuu-chah-nulth hunter does the same while hunting deer to practise *isaak*.[12] These protocols ensure a very specific relationship with the environment, people, and their corresponding spiritual importance. *Hahuulthi*, or ances-tral territory, was closely related to the practise of *isaak*. It is a way to respect an individual's inheritance of the land from their ancestors and is linked to managing it according to *isaak*.[13] It is important to under-stand that these ideologies are neither stagnant nor widespread across all Nuu-chah-nulth peoples. All societies have individuals who do not fol-low dominant social rules, and Umeek recognizes that these rules can be broken by Nuu-chah-nulth people, both in the past and present.

Umeek develops *Tsawalk* as an academic analytical tool. The pattern of *Tsawalk* demands "more rather than fewer variables"[14] during the critical examination of any study. This is in contrast to the empirical disciplines of Western academia, such as history, which tend to function by absorb-ing an acceptable amount of reliably recognized information and then producing a framework of fact that leads the audience to a final thematic conclusion. This is an alternative to accepting the notion that life, and history, abounds with more meaning than historians could ever hope to decipher. When applying *Tsawalk* to the Western discipline of history it is necessary to avoid concluding any research with definitive statements. In these situations, it becomes more important to establish the meaning of these connections rather than identify their purpose. Until Indigenous oral histories are integrated into current historical narratives of the early North-West Coast, they will always be incomplete. In the meantime, I attempt to embrace the theory of *Tsawalk* in analysing the relationships between Nuu-chah-nulth and Americans at "Adventure Cove" in the eighteenth century to further complicate, rather than simplify, the interlocking connections between culture, intention, and situation.

My goals in this essay are threefold: First I wish to use Eurocentric histories of relationships between Nuu-chah-nulth and Americans on the *Columbia* during the winter of 1792 to present the formulaic foundation of Eurocentric history. Then I hope to Indigenize the process of this history by conducting an analysis through the Nuu-chah-nulth ideology of *Tsawalk*. Finally, by not adhering to the European methods of analysis I will expand rather than diminish the possible conclusions of these circumstances. I hope to Indigenize how I have been trained to write an essay as an undergraduate student at the University of British Columbia.

European Narratives

Compiled by Frederic W. Howay in 1941, *Voyages of the Columbia to the Northwest Coast 1787–1790 and 1790–1793* included all known logs, journals, correspondence, and lists of materials that pertained to these two voyages of the *Columbia*. These accounts and letters, both received and sent from the *Columbia*, were mainly written by three men: John Boit, the fifth mate, Robert Haswell, the first mate, and John Hoskins, the clerk. These journeys began only four years after the creation of the United States of America. With the successful termination of the Revolutionary War in 1783, these newly minted Americans made first contact in the North-West Coast on behalf of the United States of America.

The sails of the *Columbia* were hoisted for the second time on 15 August 1791 when the ship left Boston full of American men, led by Captain Gray, with the instructions that included the directive "no unjust advantage taken of [Indigenous people] in trade."[15] The *Columbia*'s first visit to the North-West Coast met with financial failure. Economic success was necessary for Captain Gray on this second trip as he had purchased shares in the ship's venture.[16] Additionally, this journey warranted more national importance than the previous one as George Washington, the first president of the United States, wrote a note for Captain Robert Gray that asked foreign nations to receive the captain "upon [him] paying the usual expenses" of travel and trade.[17]

These journals only began once the Americans reached the North-West Coast, but the Clayoquot Sound area was a hotly contested political arena even before the *Columbia* arrived. By the time of contact, Chief Wickaninnish had already merged between eleven and seventeen different nations into the Tla-o-qui-aht.[18] Under his command they annihilated

(The Tla-o-qui-aht and their allies "wiped out") the Hisau'istaht Nation.[19] The Hisau'istaht's territories, trade privileges, and war allies were redistributed between the remaining three tribes (nations): the Ohqmin, the Hohpitshaht, and the Tla-o-qui-aht.[20] Opitsaht was chosen as the winter settlement for the Tla-o-qui-aht people.[21] By John Boit's account Opitsaht "contained upwards of 200 houses," making it the largest settlement on the western coast of what is now known as Vancouver Island.[22]

By September 1792 the *Columbia* had arrived on the Pacific North-West Coast as the temperatures began to drop. The ocean became a choppy, dangerous place, and Haswell suggested to Captain Gray that they should find the closest suitable cove to winter over.[23] The general plan was to construct the ship named the *Adventurer*, the material for which was placed aboard the *Columbia* before the journey began. Gray agreed, and they subsequently harboured in a small cove they fittingly founded (forcefully founded) as "Adventure Cove."[24] Gray chose this cove knowing full well that a large Tla-o-qui-aht winter village, called Opitsaht, lay three miles away.[25]

Three years before the arrival of Captain Gray, Chief Wickaninnish and the Tla-o-qui-aht began to demonstrate their military power to the people to the south in what is now known as Barclay Sound.[26] Anything from raids to complete destruction of settlements was commonplace (It was said by a Ucluelet informant that the Tla-o-qui-aht "never raided twice, but always wiped out the enemy in a single raid").[27] Weaponry that may have come from Russian, Spanish, or other European settlers aided the aggressors (the Tla-o-qui-aht) in these wars. Chief Wickaninnish requested that Haswell "allow the smiths to make daggers to kill the Highshakt people."[28] Boit also mentioned how Wickaninnish "wish'd for us to lend them musketts and ammunition."[29] To the north of Clayoquot Sound was a formidable nation, the Ahousaht of Nootka Sound. Chief Wickaninnish's tactics changed from war to diplomacy and the Tla-o-qui-aht began negotiations to etch out trade and territory agreements in that area.[30] The mediation was facilitated by marriage links between Wickaninnish and the Ahousaht chief, Tatooch.[31] There were feasts to honour the chiefs of Ahousaht when they arrived in Opitsaht.[32] Between the diplomacy in the North and military control over the South, the Tla-o-qui-aht had a monopoly over the fur trade in one of the wealthiest areas of the North-West Coast. When the *Columbia* arrived in 1972 Chief Wickaninnish controlled most relations with the American ship.

Chief Wickaninnish visited the American sailors a month after their arrival, in early October 1792. By that time, they had already finished constructing a "house" equipped with "two cannon[s] mounted" and room enough for men, arms, and food.[33] To complete their projects a "quantity of timber" needed to be cut, not to mention the need for food to cover the daily eating habits of each man.[34] A yield of "20 or 30 ducks and geese" was the result of an average day's hunt.[35] The sailors of the *Columbia* did not recognize these lands as occupied by the Tla-o-qui-aht. During a hunting expedition John Boit, the ship's fifth mate, found himself approached by Ahousaht men who shouted and waved at him. He believed that they were trying "to take his cartridge box."[36] Boit responded by pointing his musket at Chief Tatooch of Ahousaht until the men returned to their canoes.[37] Haswell agreed with Boit's assumption that their "intentions were only to rob," thereby reiterating the belief that all people of the North-West Coast are thieves.[38]

Boit began to refer to Chief Wickaninnish as "the King" after experiencing his wealthy home.[39] Yet, Captain Gray did not show similar respect for the chief. One of the few times Captain Gray visited the settlement of Opitsaht was when another chief had fallen ill, and he revived him with some Western medicines.[40] Yet this event took place in late December, three months after the Americans arrived on Tla-o-qui-aht territory. Gray never exchanged presents with the chief of the Tla-o-qui-aht, which was protocol for arriving at most Indigenous territories on the North-West Coast.[41] Also, Captain Gray was not present when John Boit witnessed Chief Wickaninnish's eldest son's naming ceremony.[42] The fact that the American leader did not show respect for the Tla-o-qui-aht leader was a cultural and political insult.

The American-Tla-o-qui-aht relationship took a turn for the worse after another event that took place on the *Columbia*. On 18 February 1792, Chief Tototeescosettle, Wickaninnish's brother, was found wearing a sailor's overcoat and was accused of theft.[43] That same day another sailor on the *Columbia* confessed that Totoocheatocose, the brother of Tototeescosettle, had propositioned him to aid the Tla-o-qui-aht in capturing the *Columbia* in return for receiving a noble standing among them.[44] Captain Gray reacted defensively and repositioned the ship away from the shore, arming all weapons both on the ship and in the fort.[45] Two days later Tototeescosettle returned with his father "to sell his skins," but Captain Gray "took the skins from him" and threatened his life.[46] In Boit's opinion these furs given to Captain Gray after the debacle of the stolen jacket was a "most specious [show] of friendship," but Gray took the furs and threatened the lives of these chiefs if they returned.[47] Within the week the Americans had stripped the "house" (their fort) of everything useful and steered the *Adventurer* and *Columbia* out of "Adventure Cove" and away from Tla-o-qui-aht territory back towards the Pacific.

The Americans escaped and nothing came of the supposed plan. However, before departing, Captain Gray ordered his sailors "to destroy the

Village of Opitsatah."[48] Hoskins and Boit both lamented the destruction of this "Work of Ages," as Boit called it,[49] but Haswell's narrative makes no mention of this order or the annihilation of Opitsaht. Gray's aggressive (murderous) reaction was not in response to a military threat, as nothing came of this apparent plan, but was, rather, a personal choice.

Intentions and Conclusions

The underlying motives shared by both Captain Gray and Robert Haswell played an important role in determining how they acted towards the Tla-o-qui-aht. Before the journey Captain Gray chose to invest in the *Columbia* and now was financially liable if it did not profit. Furthermore, the owner, Joseph Barrel, enticed them by allowing both "[Gray] five per cent... [and Haswell] one and one-half per cent" of the ship's hold for procuring their own goods.[50] A captain's portion from a prosperous mission was already enough to retire.[51] Barrel had given both the captain and first mate an incentive to make sure this was a successful voyage.

This economic incentive was paired with the order to act justly towards Indigenous peoples. Not only was there a letter from the president of the United States warning against mistreatment of Indigenous peoples, but the owner of the ship also threatened to punish a breach of contract "with the utmost severity."[52] However, it is possible that this admonishment from Barrel served as a warning for Captain Gray to avoid recording any mistreatment of Indigenous peoples in his official documents. At sea the captain held total power over the ship, its holdings, and its crew.[53]

There are many discrepancies between Haswell's log and the other accounts of the events that transpired during the *Columbia*'s six-month stay at Adventure Cove. This may be because Haswell was told by Captain Gray not to record any incriminating evidence. As second in command it was likely that Haswell's journal would comprise the official log of the *Columbia*. In contrast, Hoskins's candid statements about the events in question suggest that he was at ease criticizing Captain Gray. Hoskins's journal provides a twenty-four-page narrative of the six-month stay in Adventure Cove; by contrast, Haswell only wrote six. Furthermore, Haswell did not mention the destruction of Opitsaht, while both Boit and Hoskins did. Likewise, Captain Gray's communications with Joseph Barrel did not include anything about their encounter with Chief Wickaninnish. Hoskins's letter to Barrel was ambiguous about the entire event. He claimed they were almost finished building a house and that the *Adventurer* was nearly complete "when the natives on the 18th of February came to attack us,"[54] which was a lie as the house was finished in October of the previous year. However, Hoskins's private account of the event openly criticized Captain Gray's dealings with the Tla-o-qui-aht.[55] Hoskins could have even been told by Gray and Haswell to change his story before sending the letter.

It is clear (Multiple European accounts state) that Chief Wickaninnish was engaged in political conflict with other nations over land and trade. As a leader Wickaninnish saw that the weapons and ships of the Americans could tip the scale of war in favour of his people against those to the south or other contesting groups. Additionally, he was blocked from acquiring these tools through trade as the Americans did not allow their blacksmiths to make them weapons, nor were muskets and shot to be traded for furs. This alone may have warranted frustration, enough for Wickaninnish to plan an aggressive takeover of the ship and its property.

The resources taken by the Americans during their stay (while they invaded) Tla-o-qui-aht territory could be another reason for aggressive reactions. According to *isaak*, Nuu-chah-nulth people had a responsibility to respect the animals, plants, peoples, and spirits of the lands. Chief Tatooch of Ahousaht could have had reason to react aggressively if his *hahuulthi*, or ancestral hunting ground, was used by the Americans. When he attempted to intervene, he was threatened with violence. There were multiple levels of disrespect perpetrated by Boit: he hunted on Nuu-chah-nulth land without request; his method of hunting and harvesting may not have been in accordance with *isaak*; and Boit aggressed the owners of the area (stewards of the land) when he tried to stop them. To insult a chief and then threaten his life may have been grave enough to incur a direct attack by Chief Tatooch and the Ahousaht Nation, with or without the aid of Chief Wickaninnish and the Tla-o-qui-aht.

It is difficult to compare how the Americans and Tla-o-qui-aht practised hunting and harvesting as there were few such occasions depicted. Examples of ceremonial procedure, or *isaak*, described how the Tla-o-qui-aht "rub their face with a piece of [snake]" before they embarked on a whale hunt.[56] Another example was that when Tla-o-qui-aht people brought wood to the Americans it was "split with wedges from the Log."[57] This practice does not require the tree to be cut down or stripped of its limbs. In stark contrast, the image of Adventure Cove showed the trees stripped of all their limbs, assuredly by the Americans. This was one example of how the American methods of collecting resources differed from the Tla-o-qui-aht. However, within a Nuu-chah-nulth perspective the two transgressions, stealing and harvesting without practising *isaak*, were inseparable according to *Tsawalk* as all relationships are interconnected.

It could be assumed that a chief of such affluence and authority as Wickaninnish would respond negatively to insult and inconvenience.

However, there was no evidence that Chief Wickaninnish or the Tla-o-qui-aht were ever involved in the planned attack on the *Columbia*, or indeed that there was an attack at all. Noble marriages have been used by all cultures to unite factions and gain important allies. It was as likely that the brothers of Wickaninnish were chiefs from other nations and that they had entirely different intentions than those pursued by Wickaninnish. These factions could have acted separately or together with Chief Wickaninnish, or possibly in total secrecy and disobedience. The politics of any society are difficult to navigate, and the politics of the North-West Coast were no different.

If all relationships were important to the Nuu-chah-nulth, under the belief of *Tsawalk*, then the visitors who interacted with their lands, resources, and people were equally significant as their own. The Tla-o-qui-aht could have considered the Americans' actions as theft but permitted it. The Americans, in essence, were a part of *Tsawalk*, and just like the Tla-o-qui-aht, they needed the warmth, sustenance, and protection that the cove provided. In the end the possibility of an attack may have resulted from Captain Gray threatening the life of Chief Tototeescosettle, and had less to do with the taking of any resources.

What is important to note is that the American narratives did not depict a single successful theft perpetrated by the Tla-o-qui-aht <u>over the entirety of their stay</u> (while they squatted on their land). Chief Tatooch's guards were not able to take John Boit's box of munitions, the sailor's overcoat was returned, and the plan to take over the *Columbia* and the *Adventure* failed. Yet, failed attempts at "thefts" performed by the Tla-o-qui-aht were more likely to be recorded in the journals of the *Columbia* (as opposed to the repeated and actual theft of Tla-o-qui-aht resources by the Americans). By consistently depicting the Tla-o-qui-aht as thieves in these narratives, the Americans positioned themselves as upright and moral. This not only degraded the representation of Indigenous people in Western society but encouraged the misconduct of the sailors of the *Columbia* to continue. The *Voyage of the Columbia* was just one beginning of the appropriation and misrepresentation of Indigenous history in the North-West Coast through written language.

The reactions of John Boit and Captain Gray to the "theft" of a munitions belt and overcoat were exemplary of their belief in private property. Yet, upon closer analysis this belief in ownership was only recognized by the Americans when European powers were involved. The cove that the *Columbia* wintered in was near Opitsaht but was not considered owned by the Tla-o-qui-aht people. However, the Americans were commanded not even to touch the Spanish settlement of Nootka Sound.[58] Joseph Barrel had never been to the North-West Coast, but he insinuated Indigenous ownership of land when he told Captain Gray to purchase the land from them. Economic benefit fuelled the American decision to ignore Indigenous peoples' ownership, but the underlying Western assumption (racist assumption) that Indigenous peoples were less than Europeans would have presumably played an important role. Either way, it was unlikely the Americans were ignorant of Tla-o-qui-aht ownership since the crew of the *Columbia* went to such ends to conceal the events during the winter of 1792 to avoid punishment from Joseph Barrel.

It was made clear to Captain Gray and Haswell that a successful voyage meant significant wealth for both. The enticement was enough for them to falsely depict the Tla-o-qui-aht as thieves, intentionally and unintentionally steal from them, and then lie about the encounters upon their return. Was this purely to become rich, to slight Indigenous peoples, or both? While men like Captain Gray did not practise *Tsawalk* it was likely that his foul deeds had a relationship with honourable notions (resulted from more than just greed). The life of a sailor was dangerous, unpredictable, and took a person away from their family for years at a time. The promise of wealth extended by Barrel could have been enough for Captain Gray to become a businessman at home with his wife and four daughters.[59] While this may be admirable in a way, it was based primarily on selfish individualism and racist tendencies. These Americans committed horrible (These Americans committed murderous) actions towards the Tla-o-qui-aht people in 1792, and these actions continue to be perpetuated to this day (by the settler state. More recently, for example, three Tla-o-qui-aht members were fatally shot by police within a span of eleven months, including Chantel Moore in 2020 and Julian Jones in 2021; these followed the shooting of a Tla-o-qui-aht mother in 2011. It is clear that the settler-colonial violence against Indigenous peoples inherent in the creation of Canada and the United States is still present in these countries' systems today).[60]

I have not completely succeeded in analysing these Eurocentric histories through the Nuu-chah-nulth world view of *Tsawalk*; I did not expect to, but I have attempted to complicate a privileged telling of the events at Adventure Cove during the winter of 1792–93. The histories of the world have been reshaped and redeveloped by the same Western peoples and through the same Western methodologies. To truly begin to unpack the full history of early contact on the Pacific edge of Turtle Island, an Indigenous oral tradition must be integrated. History should no longer be a monolith of Western evidence that obscures the multiplicity of a people's history (complex histories of Indigenous communities throughout the invasion and colonization of Turtle Island).

NOTES

1 Richard E. (Umeek) Atleo, *Tsawalk: A Nuu-chah-nulth Worldview* (Vancouver: UBC Press, 2004), 125.
2 Umeek (Atleo), *Tsawalk.*
3 Nuu-chah-nulth Tribal Council, accessed October 22, 2024, http://www.nuuchahnulth.org/.
4 Frederic W. Howay, *Voyages of the "Columbia"to the Northwest Coast 1787–1790 and 1790–1793* (New York: Da Capo Press, 1941).
5 Howay, *Voyages of the "Columbia,"* 33.
6 Philip Drucker, *Indians of North America: Northwest Coast of North America* (New York: Natural History Press, 1963), 30.
7 Daniel W. Clayton, *Islands of Truth: The Imperial Fashioning of Vancouver Island* (Vancouver: UBC Press, 2000), 131.
8 Umeek (Atleo), *Tsawalk.*
9 Umeek (Atleo), 125.
10 Umeek (Atleo).
11 Umeek (Atleo), 130.
12 Umeek (Atleo).
13 Umeek (Atleo), 127.
14 Umeek (Atleo), 118.
15 Howay, *Voyages of the "Columbia,"* vii.
16 Howay, 307.
17 Howay, 444.
18 Clayton, *Islands of Truth*, 131.
19 Clayton, 132.
20 Clayton.
21 Clayton.
22 Howay, *Voyages of the "Columbia,"* 391.
23 Howay.
24 Howay, 252.
25 Howay, 303.
26 Clayton, *Islands of Truth*, 143.
27 Clayton, 144.
28 Howay, *Voyages of the "Columbia,"* 313.
29 Howay, 389.
30 Clayton, *Islands of Truth*, 143.
31 Clayton, 143.
32 Clayton, 135.
33 Howay, *Voyages of the "Columbia,"* 444.
34 Howay, 252.
35 Howay, 305.

36 Howay, 307.
37 Howay, 306.
38 Howay.
39 Howay, 442.
40 Howay.
41 Drucker, *Indians of North America*, 30.
42 Howay, *Voyages of the "Columbia,"* 310.
43 Howay, 312.
44 Howay, 275.
45 Clayton, *Islands of Truth*, 132.
46 Howay, *Voyages of the "Columbia,"* 385.
47 Howay, 384.
48 Howay, 383.
49 Howay, 386.
50 Howay, 390.
51 Drucker, *Indians of North America*, 30.
52 Howay, *Voyages of the "Columbia,"* 445.
53 Daniel Vickers and Vince Walsh, *Young Men and the Sea: Yankee Seafarers in the Age of Sail* (New Haven, CT: Yale University Press, 2005).
54 Howay, *Voyages of the "Columbia,"* 483.
55 Howay, 475.
56 Howay, 383.
57 Howay, 382.
58 Howay, 380.
59 Howay, xiv.
60 "RCMP Shoot Tla-o-qui-aht Woman on Mother's Day Weekend: Indigenous Leadership Furious with Ongoing Police Brutality," Union of BC Indian Chiefs, 11 May 2021, https://www.ubcic.bc.ca/mother_s_day_weekend _indigenous_leadership_furious_with_ongoing_police_brutality.

Chapter Ten

The Fight for Water: Examining Environmental Racism and the Effects on First Nations Culture and Society in British Columbia

KEVIN LY

Figure 10.1. Photo taken at Edith Lake, in the territory of the Skwxwú7mesh-ulh Temíxw (Squamish), Cayuse, Umatulla, and Walla Walla Nations, 1 January 2020.

Source: Courtesy Kevin Ly.

Introduction

Indigenous communities in British Columbia have always viewed water as a substantial element in the creation of life. In Michael Blackstock's interview with elders from the Secwepemc, Nlaka'pamux, and Syilx First Nations, water was described as "Mother Earth's blood," a "meditative medium," and a source of nourishment for medicinal plants, and is thus thought to possess a power that requires respect.[1] However, as of March 2021, 38 First Nations communities across Canada were subject to 58 different long-term drinking water advisories, which are alerts by the government that waterborne contaminants are present in local drinking water sources.[2] Despite a decrease in the initial count of 101 drinking water advisories in 2015, this continued lack of access to clean water represents not only a dangerous health risk to Indigenous communities but also a denial of First Nations' access to cultural and social practices involving water; in effect, this neglect by the Governments of Canada and British Columbia is a clear example of environmental racism. This essay will first examine the connection that First Nations in British Columbia have to water (physical and metaphysical) before looking at how uneven access to clean water in these Indigenous communities provides substantial evidence of environmental racism. Finally, and more substantially, this essay will argue that this granting of uneven access by the government results in discriminatory action towards Indigenous communities, and that by denying Indigenous communities access to clean, safe water, the government is also creating harmful impacts on First Nations' cultural, social, and political practices.

British Columbia First Nations and Water

Water is undeniably an important part of the physical environment, especially within Canada; as one of the most common substances on earth, it is essential to survival for many organisms within our ecosystem. However, water has also long been a critical component of culture and is a significant influence on societies. Indigenous communities in British Columbia are no exception; not only are they impacted directly by water (and lack of access to it), but many Indigenous communities also see water as significant in a number of other respects.

On a material and physical level, the significance of water can be seen in the livelihoods of the Coast Salish peoples. The Coast Salish relied heavily on the ocean, and as such viewed the waters of the Pacific with reverence. Not only did they use the Salish Sea as a source of nutrition and as hunting grounds for seafood (such as seaweed, crab, seal, salmon, herring, and candlefish, known as *oolichan*), but they also performed ceremonies on the waters before long fishing expeditions. The Coast Salish believed the ocean contained spirits that could either harm or protect fishers.[3] Coast Salish peoples (along with the Nuu-chah-nulth and Kwakwakaw'akw) also built their homes to face the ocean, due to their recognition of the waters as both food source and avenue of transportation. In northern British Columbia, the Tsimshian, Tlingit, and Nisga'a Nations also built their villages and lives around water sources, due to their reliance on marine animals for food; indeed, their legends often revolve around marine animals, as well as the rivers and watersheds within their territory, in a display of respect for water. (A Nisga'a legend holds that volcanoes within the area erupt because young boys were abusing a salmon to explain the young volcanoes that exist within the geography of the Nass River Valley.) Other tribes, such as the Dunne-za, derive their names from the surrounding waters and the animals that reside in these aquatic environments; roughly translated, the Dunne-za are known as "dwellers among the beavers" due to their proximity to the Peace River.[4] The First Nations' respect for water and those creatures that reside in the aquatic environment is also reflected in the spiritual and symbolic significance accorded these beings.

The First Nations of British Columbia perceive water as a biotic component – that is, a living entity. This observation of water as a spirit translates into different perceptions of water in traditional ecological knowledge. British Columbia's physical geography and its location by the Pacific Ocean also makes many First Nations' connections to water unique relative to other First Nations throughout Canada. For many British Columbia First Nations, water is also regarded as being a "meditative medium" and "purifier"; it is central to many First Nations rituals. For example, the Syilx Nation regard water as a powerful spirit, one that can relieve stress and provide calm. Secwepemc medicine men were known to swim "morning and night when they were practicing their medicine powers."[5] Springs, as a source of water for plants cherished by First Nations, are also sources of water for villages and lodges, proved by the existence of settlements such as a Stl'al'imc pithouse village by Lillooet, BC.

The sweat lodge, perhaps one of the most prominent cultural features found within many First Nations, both inside and outside the province, is the traditional site of a purification ritual that, similar to immersion in water, is meant to restore the body to its purest form; water, as steam, is a spiritual cleanser and expeller in this ceremony, and the location of the sweat lodge by rivers, lakes, or springs represents a physical connection to the landscape (the purification aspect of the sweat lodge will be explored further on in this essay.) Within the lodge itself, water is poured onto hot rocks, and the steam is meant to help the individual sweat out and essentially rid their body of their ills. Evidence for the healing properties of this ritual can be found in the present day, outside Indigenous communities, with modern applications of the sweat lodge found in holistic medicine to treat such things as addiction.[6] The sweat lodge also symbolizes the womb of Mother Earth: the emergence of the individual from the sweat lodge symbolizes a rebirth, and in certain cultures, the individual then bathes themselves in the source of water (a lake, river, stream, or the ocean) nearby.

Ecologically, water is also perceived by First Nations as helping maintain a balance in the natural environment. Groundwater is heavily emphasized as being crucial; First Nations believe that the trees and vegetation act as pumps, dispersing groundwater and storing it within the forest.[7] The ecological knowledge implicit within this understanding of groundwater and water pumps is reflected in the modern concept of the "water table." Indigenous elders also refer to this balance when discussing the evaporation of nearby creeks and rivers due to logging further upstream; Blackstock reports that from a contemporary scientific view, these water tables are influenced by flows in the rivers, the presence of deciduous vegetation, and reduced fog interception.[8] Therefore, these physical observations further emphasize that water is heavily imbued in understandings of traditional ecological knowledge.

First Nations' Water Rights

Following the understanding developed by First Nations in Ontario in 2008 in their Water Declaration, I consider "water" in this essay to include "rain waters, waterfalls, rivers, streams, creeks, lakes, mountain springs, swamp spring, bedrock water veins, snow, oceans, icebergs and the seas."[9] The recognition of Indigenous water rights is evidently a very detailed affair, one that has long been a contentious issue in Canada, as evidenced by the lack of clean water access available to First Nations communities and reserves.

As stated above, as of March 2021, there were thirty-eight Indigenous communities under a drinking water advisory, despite a promise by the federal government to end the water crisis by 2021; in many cases, residents are unable to drink water without first boiling it, and sometimes water is not fit for consumption or bathing. A Human Rights Watch report in 2016 found contaminants such as E. coli, uranium, and carcinogenic trihalomethanes.[10] Less than a decade ago, British Columbia had the second-highest number of boil-water advisories behind Ontario, and the highest rate of waterborne illness due to municipalities utilizing surface water, with some communities only recently lifting long-term drinking water advisories lasting as long as sixteen years.[11]

However, these advisories only acted as a Band-Aid solution to the problems. According to the Canadian Medical Association in 2012, it would collectively cost $31 billion to upgrade water and waste-water treatment infrastructure in Canada.[12] In 2005, the federal commissioner of the environment and sustainable development declared that, "despite the hundreds of millions in federal funds invested, a significant proportion of drinking water systems in First Nations communities continue to deliver drinking water whose quality or safety is at risk," and that "residents of First Nations communities do not benefit from a level of protection comparable to those of people who live off reserves."[13]

One may ask, Who is responsible for managing drinking water, and why has it taken so long for the Canadian government to act? Under Canadian law, chiefs and councils "are responsible for planning and ... the day-to-day operation of water and wastewater systems on reserves,"[14] while Indigenous Services Canada (formerly Indian and Northern Affairs Canada) provides "funding for water services and infrastructures."[15] However, questions concerning the management of drinking water are moot when there is little legal framework guiding who provides water to communities. Phare points out three logistical issues behind the current management and provision system to First Nations communities.[16] First, the *Indian Act* fails to define comprehensive and clear powers related to water management. Second, many provincial laws do not apply on reserves; water protection laws therefore are of little use within First Nation communities, and legislative authority for the provision of drinking water to reserves lies with the federal government. And third, the lack of a comprehensive legal standard for the provision of drinking water means each province creates its own "patchwork of standards and rules that apply within their boundaries"[17] – the national *Water Act*, for instance, has not been updated since the 1970s.[18] These three facts together describe the challenges that face policymakers and First Nations leaders who campaign for the right to water.

These political and logistical hindrances place First Nation communities at a higher risk of waterborne diseases compared to the general population. Here we see matters of environmental justice and racism reflected in the inability of the government to provide First Nations communities with access to clean water. In October 2005, this environmental injustice was made evident with the evacuation of over a thousand residents from the Kasheshwan reserve in northern Ontario after E. coli was discovered in the water. In response to this event, the commissioner of the environment and sustainable development identified that "most treatment plant operators do not possess the knowledge and skills required to operate their plants safely," and confirmed Phare's identification of key logistical challenges by noting that the "technical help available to First Nations to support and develop their capacity to deliver safe drinking water is fragmented."[19]

The delivery of safe drinking water to First Nations reserves was noted by the Government of British Columbia to be "critical to the health and safety of the communities' residents."[20] Besides this, the provincial government also noted that access to clean water is also part of the "economic viability" that these communities currently lack.[21] However, it is evident that the lack of access to clean water in First Nation communities is not simply a matter of economics; it is a critical social issue that speaks to the priorities of the British Columbia government. In fact, this obvious lack of access for First Nations communities speaks to a greater problem, one that extends beyond health problems, political bureaucracy, and human rights. The delivery of clean water (or lack thereof) is classified by many critical geographers as an issue of environmental racism.

What Is Environmental Racism, and Where Is the Evidence?

The term "environmental racism" describes the deliberate placement of communities in less desirable locations (characterized by the presence of such things as hazardous waste sites, landfills, incinerators), and/or, as in this case, the location of communities in areas excluded from the mainstream, dominant culture. This intentional placement and marginalization denies full citizenship to these communities, and usually leaves members with little political power and representation in the policymaking process.[22] This marginalized status defines certain reserves all too well; as Mascarenhas writes, "the remote nature of many First Nations communities joined with the special jurisdictional issue associated with them, has led to a lack of clear responsibilities for the health of these communities."[23]

Environmental racism is also exemplified in cases "where the dominant culture perceives subordinated others as a 'resource' with no goals and purposes of their own," and "where the subordinated other [First Nations communities] is defined solely in terms of the dominant culture."[24] First Nations settlements constitute a high proportion of those communities affected by water advisories, and the number has grown by nearly 40 per cent since 2006.[25] Some communities have been on water advisories for sixteen years, with some residents lacking running water in their homes. Since 2011, thirty-one British Columbia communities have been under water advisories, and many communities remain overcrowded. A number are still unable to use their water, even after boiling, due to the risk of gasoline and trihalomethane (a chemical compound that is related to a higher risk of cancer).[26] It is worth comparing the official response to this phenomenon to the treatment of the Walkerton incident, where the provincial and municipal governments confirmed the presence of E. coli within a month. In the words of Chief Moonias of the Neskantaga Nation,

> I am of the opinion that when Walkerton, Ontario faced its water crises a few years ago, the Provincial Government did not respond to this municipality that it was an operations and maintenance issue or only offered to assist by providing the community with an advance of funds. This is the current position of the Department of Indian and Northern Affairs Canada. I wonder how different the response would be if the residents of Toronto were without access to water?[27]

The Canadian government has also abstained from choices that would enable First Nations to have clean access to water. Internationally, the United Nations General Assembly recognizes the right of every citizen to "safe and clean drinking water and sanitation" (in a vote that Canada conspicuously abstained from).[28] In 2008, the Canadian government backed away from the Kelowna Accord, an agreement that would've dedicated $5.1 billion to "improving the socio-economic conditions and access to water for Aboriginal people."[29] While the government did take some action (it set aside $330 million in the 2008 budget), First Nations leaders were deeply disappointed with the choice. The Conservative government then voted against Bill C-292, which had been implemented to recognize the Kelowna Accord, even after it announced that the $330 million was not enough to "ensure safe drinking water in all First Nations communities when considering the need for new or upgraded infrastructure."[30] This evident lack of political commitment is confirmation of Draper and Mitchell's observation that "in Canada, relatively little policy discussion explicitly linked to environmental justice has occurred ... political and private sector leaders in Canada generally have not taken strong positions related to environmental justice issues."[31]

This position has been contrasted with the commitments made by the Trudeau government: between 2016 and 2023, the Trudeau government has "made over $5.6 billion in commitments to First Nations to upgrade water and wastewater infrastructure on reserve."[32] However, the announcement by the same government to invest almost $13 billion to build the state-owned Trans Mountain pipeline (purchased from Kinder Morgan in 2018 for $4.5 billion) has rendered these prior commitments moot: potential oil spoils, pegged at 79 to 87 per cent certainty by the Tsleil-Waututh Nation Sacred Trust initiative, threatens not only the Burrard Inlet and the Fraser River, but also the Coldwater Valley watershed and Coldwater aquifers (vital for the Coldwater Indian Band), among other water sources.[33]

John Borrows, in "Living between Water and Rocks: First Nations, Environmental Planning and Democracy," introduces the idea that First Nations are not only living at the geographical margins of the land but also "exist just beyond the borders of the North American legal imagination."[34] The limits that Canadian First Nations face in environmental planning render them "invisible" and "repressed" under the federal structures of the Canadian state. However, this repression extends beyond the political arena; as we will see, the environmental racism that First Nations in British Columbia face in their access to water also has an impact on their cultural and social practices.

The Impact of Environmental Racism on First Nations Communities

First Nations' lack of access to clean water not only reflects a grave disregard for human rights, public health, and the effects of environmental racism; it also has a powerful impact on many facets of Indigenous life. These problems are clearly not solely physical or environmental in nature; as will be discussed, the lack of access to water also has profound spiritual and cultural impacts on First Nations, and also on First Nations' notions of identity.

British Columbia First Nations, as detailed above, perceive water as a life-giving substance ("the lifeblood of Mother Earth"); its life-sustaining properties have helped define intrinsic and complex relationships found within the cultures of different First Nations in the province.[35] First Nations' cultural identity therefore is tied to, and, more crucially, threatened by, the lack of access to clean water. In Michael Blackstock's *Water: A First Nations' Spiritual and Ecological Perspective*, First Nations elders were questioned about the significance of the degradation of water quality to their present way of life. Most notably, one elder noted that "when you start depending on yourself to survive without spirituality – the thanksgiving, thanking Mother Nature, the creator for their gifts – you'll suffer because of it ... Mother Nature has its way of disciplining us."[36] Another elder also spoke about the impact of water quality on culture: water rituals such as *amo:hi atsv:sdi* (meaning "water place, to go and return to, one") were important for newborns. "The water is the biggest part of all our lives; without it we'd never survive ... with the first born, [the

Secwepemc] take their babies to the water and dunk that baby into the water. It's steaming, gives that baby strength; it shares its life with that baby, its energy."[37]

Along with these cultural acts, scholars have also detailed the meditative properties of water to British Columbia First Nations. The healing properties of water are what Kathleen Wilson terms "therapeutic landscapes"; because Indigenous peoples regard their physical landscapes as shaping their "cultural, spiritual, emotional, physical and social lives," Wilson contends that physical landscapes can act as "locations of healing" within First Nations societies.[38] Research has dictated that symbolic structures within such therapeutic landscapes transcend notions of physical space and place; that is, therapeutic landscapes are not necessarily physical in nature, but are also "embedded within the belief and value systems of different cultural groups."[39]

The healing properties of rituals involving water, such as the sweat lodge, are multifaceted; besides the symbolism of the connection between the individual, Mother Earth, and the Creator, these rituals also represent the importance of symbols in shaping health within Indigenous communities. James Waldram contends that health among Indigenous populations is not simply based on the physical removal of disease, but must also contain notions of symbolic health, "dependent upon the use, interpretation, negotiation, and manipulation of cultural symbols as central to the process of healing."[40] The removal of access to water is therefore also a removal of a physical landscape that is central to notions of healing within Indigenous communities. As a part of Indigenous culture and religion, water is also central to certain notions of Indigenous cultural identity. The restriction of Indigenous communities inflicted by the environmental racism implicit in policy and practice therefore inhibits First Nations from fully participating in healing and cultural rituals and engaging in their conceptions of health. However, the lack of access to water also disrupts notions of identity and expression.

Wilson and Peters found that in some First Nations' concepts of identity, landscape features such as water were a significant part of expressing and understanding one's relationship to the land.[41] The absence of such features made respondents "uncomfortable": "I can go outside, take my tobacco outside everyday and lay it by a tree … I improvise … I tell [the elders of the community] I really miss the bush or um I miss being home, they say 'well go to water …' and that helps me but it's not the same."[42] These improvised spaces, however, were found to be bereft of the same connection. Another respondent noted that their community had "really lived off the land. I tend to feel a better connection [when] I am by the water, the rocks and the tree and all the islands"; the reserves also, in comparison to cities and urban settings, represent "an important source of cultural identities and cultural practices."[43] Therefore, the lack of access to clean water and the removal or destruction of natural features that once provided clean water to communities has a significant impact on the social identities of First Nations community members. Traditional ecological knowledge (TEK) is also affected by the absence of water in a natural environment; as detailed previously, the observation of rivers and streams can offer indications of the health of a forest and activities that influence the biotic environment.

However, the inability to access clean water may also inhibit the transmission of TEK to new generations of Indigenous people. The "rediscovery" and integration of TEK has benefited watershed management in Manitoba, for example. Efficient but adaptive natural resource management is also critical to economic progress, especially in British Columbia. It is evident that local and traditional understandings of the environment, accumulated incrementally and "tested by trial-and-error and transmitted to future generations," are at risk of being lost if future generations are denied access to these same natural resources, be it as a result of drinking water advisories or potential oil spills from federally mandated pipelines.[44] TEK also offers an alternative to Western scientific ecological knowledge in that it does not aim to harness or control nature; its emphasis on relations, conservation, and its alignment with environmental planning are evidently useful for modern resource management.

Conclusion

The lack of access to clean water in Indigenous communities, combined now with the continued threat of oil spills in water sources, is nothing short of a human rights violation. The disproportionate number of Indigenous communities throughout Canada without access to clean water, and the presence of British Columbia First Nations reserves among those communities, is a warning sign of the lack of environmental justice. The lack of policy, along with the plethora of political bureaucracy and judicial red tape, are some of the reasons behind why Indigenous communities continue to face issues regarding their water supplies. There is also poignant evidence of environmental racism found in the marginalization of these same communities, and continuing lack of action from the government, at both the provincial and federal levels, especially with regards to federal economic interests. As this essay has noted, environmental racism has a significant health impact on these communities.

However, as this essay has also argued, the lack of water rights for Indigenous communities in British Columbia effects the important and complex relationships these communities have with their physical landscapes, and more specifically with water. In threatening the cultural rites and rituals surrounding water, the symbolic importance and respect with which First Nations imbue water, there is also an enduring negative effect on the cultural processes, social identities, and personal expressions that are tied to water. It is therefore evident that the unique and multifaceted relationships that British Columbia First Nations have with water and their physical landscapes do not only complicate discussions of environmental racism; they also elevate the question of water access to a human rights issue that now has profound consequences for culture, identity, and expression.

NOTES

1 Michael Blackstock, "Water: A First Nations' Spiritual and Ecological Perspective," *Journal of Ecosystems and Management* 2, no. 1 (2001): 62.
2 "Progress Update on the Government of Canada's Commitment to Clean Water in First Nations Communities," Indigenous Services Canada, 10 March 2021, https://www.canada.ca/en/indigenous-services-canada /news/2021/03/progress-update-on-the-government-of-canadas -commitment-to-clean-water-in-first-nations-communities.html.
3 Rudolph C. Ryser and Leslie Korn, "A Salish Feast: Ancient Roots and Modern Applications," *Cultural Survival*, 26 May 2010, https://www .culturalsurvival.org/publications/cultural-survival-quarterly /salish-feast-ancient-roots-and-modern-applications.
4 Fasken Martineau LLP, *Site C Clean Energy Project*, vol. 5, appendix A03, part 1, *Community Summary: Blueberry River First Nations* (Vancouver: Site C First Nations Engagement Team, January 2013), 17, https://iaac-aeic.gc.ca/050 /documents_staticpost/63919/85328/Vol5_Appendix-Blueberry_River.pdf.
5 Blackstock, "Water," 58.
6 Kathi Wilson and Evelyn J. Peters, "'You Can Make a Place for It': Remapping Urban First Nations Spaces of Identity," *Environment and Planning D: Society and Space* 23, no. 3 (2005): 395–413; Kathleen Wilson, "Therapeutic Landscapes and First Nations Peoples: An Exploration of Culture, Health and Place," *Health and Place* 9, no. 2 (2003): 89–93.
7 Blackstock, "Water."
8 Blackstock.
9 Merrel-Ann S. Phare, *Denying the Source: The Crisis of First Nations Water Rights* (Surrey, BC: Rocky Mountain Books, 2010), xi.
10 "Make It Safe: Canada's Obligation to End the First Nations Water Crisis," Human Rights Watch, 7 June 2016, https://www.hrw.org /report/2016/06/07/make-it-safe/canadas-obligation-end-first -nations-water-crisis.

11 Harpa Isfeld, *Boil Water Advisory Mapping Project: An Exploration and Review of Available Data* (Winnipeg: Prairie Women's Health Centre of Excellent, 2009); Indigenous Services Canada, "Progress Update."

12 Laura Eggertson, "Investigative Report: 1766 Boil-Water Advisories Now in Place across Canada," *Canadian Medical Association Journal* 178, no. 10 (2008): 1261–3.

13 *Report of the Commissioner of the Environment and Sustainable Development to the House of Commons*, chapter 5, "Drinking Water in First Nations Communities" (Ottawa: Office of the Auditor General of Canada, 2005), 2, https://www.oag-bvg.gc.ca/internet/docs/c20050905ce.pdf; Phare, *Denying the Source*, 9.

14 Jason G. Nang, "Developing an Environmental Management Plan for the Bras D'or Lakes Watershed: An Analysis of Its Scope and Approach for Addressing Issues" (master's thesis, Dalhousie University, 2007), 131, https://waves-vagues.dfo-mpo.gc.ca/library-bibliotheque/331508.pdf.

15 Nang, "Developing an Environmental Management Plan," 131.

16 Phare, *Denying the Source*, 13.

17 Phare, 13.

18 Eggerston, "Investigative Report."

19 Tonina Simeone, *Safe Drinking Water in First Nations Communities* (Ottawa: Library of Parliament, 2010), 5.

20 Simeone, *Safe Drinking Water*, 11.

21 Simeone.

22 Michael Mascarenhas, "Where the Waters Divide: First Nations, Tainted Water and Environmental Justice in Canada," *Local Environment: The International Journal of Justice and Sustainability* 12, no. 6 (2007): 565–77.

23 Cited in Christina Dhillon and Michael G. Young, "Environmental Racism and First Nations: A Call for Socially Just Public Policy Development," *Canadian Journal of Humanities and Social Sciences* 1, no. 1 (2010): 25–39.

24 Greta Gaard, "Women, Water, Energy: An Ecofeminist Approach," *Organization and Environment* 14, no. 2 (June 2001): 162.

25 Mark Blackburn, "More First Nations under Drinking Water Advisories," *APTN News*, 3 February 2012, https://www.aptnnews.ca/national-news /more-first-nations-under-drinking-water-advisories/.

26 Andrea Harden and Holly Levaillant, *Boiling Point: Six Community Profiles of the Water Crisis Facing First Nations within Canada* (Ottawa: Polaris Institute, 2008), 9.

27 Harden and Levaillant, *Boiling Point*, 9.

28 "General Assembly Declares Access to Clean Water and Sanitation Is a Human Right," UN News Centre, 28 July 2010, https://news.un.org/en /story/2010/07/346122.

29 Harden and Levaillant, *Boiling Point*, 8.

30 Harden and Levaillant, 8.

31 Cited in Michael Buzzelli, *Environmental Justice in Canada: It Matters Where You Live* (Ottawa: Canadian Policy Research Networks, 2008), 10, https://oaresource.library.carleton.ca/cprn/50875_en.pdf.

32 "Appearance before the Standing Committee on Public Accounts on OAG Report 3: Access to Safe Drinking Water in First Nations Communities, November 30th, 2023," Government of Canada, last modified 6 March 2024, https://www.sac-isc.gc.ca/eng/1706039365450/1706039383699.

33 "Would the Proposed Trans Mountain Pipeline Risk British Columbia Drinking Water?," Tsleil-Waututh Nation Sacred Trust, accessed 27 November 2023, https://twnsacredtrust.ca/concerns/drinking-water/.

34 John Borrows, "Living between Water and Rocks: First Nations, Environmental Planning and Democracy," *University of Toronto Law Journal* 47, no. 4 (1997): 418.

35 Mascarenhas, "Where the Waters Divide."

36 Blackstock, "Water," 61.

37 Blackstock, 58.

38 Wilson, "Therapeutic Landscapes and First Nations Peoples."

39 Wilson.

40 Wilson.

41 Wilson and Peters, "You Can Make a Place for It."

42 Wilson and Peters.

43 Wilson and Peters.

44 Fikret Berkes, Johan Colding, and Carl Folke, "Rediscovery of Traditional Ecological Knowledge as Adaptive Management," *Ecological Applications* 10, no. 5 (2000): 1252–62.

Chapter Eleven

Indigenous Legal Systems and the Struggle for Recognition

TOSIN FATOYINBO

Figure 11.1. Photo taken at Lake Louise, 2021.

Source: Courtesy of Tosin Fatoyinbo.

Written is more reliable than spoken. Getting there second means you have first right to it. Books are more informative than dreams. Singing is not relevant. Land is idle. When newcomer peoples take land it is rightful, before we can fulfil our obligations and remind you of our relationship with it, you label it a claim and insist we prove it. When Canadian judges assert that sovereignty crystallized, it is not fantastic, it is reality and cannot be argued. When Indigenous peoples and advocates assert that our sovereignty is inherent and ongoing as are our rights to land, that is fanciful and leads to an argument. Newcomer rights are presumed and Indigenous rights must be proven. Who is the savage? It is barbarism. This land is fertile because our ancestors and relations were here thousands of years and became that earth. Surely that is some measure of rightful belonging. We belong to that land.[1]

Introduction

Legal systems and traditions existed and were practised by Indigenous nations in Canada prior to the arrival of the colonizers.[2] These nations included, among many others, the Assiniboine, Dakota, Gitksan, Innu, Mi'kmaq, Cree, Maliseet, Montagnais, Saulteaux, Dene, Blackfoot, Haudenosaunee, Lakota, Nagoya, Inuit, Métis, Anishinabek, and Haida.[3] They had rich laws, traditions, customs, cultures, and values that defined the nature of relationships within their nations.

For many generations, the Government of Canada targeted the legal, cultural, and political systems of Indigenous peoples. Indigenous peoples and traditions were viewed as barbaric; their cultures and religions subjugated; their languages described as crude and repressed; and their rights undermined and denied.[4] The overall aim of these acts of oppression was genocidal with the intent to eliminate Indigeneity and absorb Indigenous peoples into the Canadian body politic.[5] It was part of a targeted policy of assimilation that employed methods of intimidation aimed at diminishing the self-worth and self-belief of Indigenous people in the eventual hope of eliminating them culturally and physically.[6] Despite the many assaults on these traditions, they continue to stand. Damages were undeniably done to the traditional and religious architecture; however, it is a credit to the strength of Indigenous people that they have bravely withstood the legacies of colonialism.[7]

Indigenous laws and legal systems also became casualties of colonialism and Crown sovereignty and supremacy.[8] Understanding that governance or a system of governance is central to any legal system, "Canada replaced existing forms of Aboriginal government with relatively powerless councils whose decisions it could override and whose leaders it could depose."[9] As a result, Indigenous legal systems have been entangled in an asymmetrical struggle with the colonial legal system that had become pre-eminent due to years of subjugation, oppression, and suppression; however, there has been a resurgence in discussions of the vital role of Indigenous legal systems in the self-governing apparatus of Indigenous peoples.[10] The Truth and Reconciliation Commission of Canada identified the recognition and implementation of Indigenous laws and systems as follows: "Aboriginal peoples must be recognised as possessing the responsibility, authority, and capability to address their disagreements by making laws within their communities. This is necessary to facilitate truth and reconciliation within Aboriginal societies."[11]

This essay analyses the struggles between the two legal systems, and the subjugation of Indigenous legal systems when they threaten the non-Indigenous legal system (i.e., Canadian laws). To do this, I review the early 2020 conflict between the hereditary chiefs of the Wet'suwet'en First Nation and the Coastal GasLink project in northern British Columbia. In particular, I explore the tension between the elected band council and the hereditary chiefs, and how corporations like Coastal GasLink exploit elected band councils to further the agenda of the non-Indigenous legal system.

This essay is divided into five parts. The first section introduces the research subject and clarifies the intention and objectives of the essay. The second section reviews available literatures on Indigenous and non-Indigenous legal systems, including their origins and features. A discussion on the relationship between both systems forms the core of the third section. That and the subsequent section review the Indigenous legal system of the Wet'suwet'en First Nation, outline the traditional and historic roles of the hereditary chiefs, and delve into the tension between the hereditary chiefs and elected chiefs in the context of the Coastal Gas-Link pipeline project in northern British Columbia. The fifth and final section then offers some concluding thoughts.

Non-Indigenous versus Indigenous Legal Systems

This section traces the origin and key features of both non-Indigenous and Indigenous legal systems. The term "non-Indigenous legal system" in this essay refers to Canadian laws/legal systems and also includes English and French law traditions. While acknowledging the fact that customs and traditions differ in different Indigenous nations, the term "Indigenous legal systems" is cautiously used to refer to the conceptualization of the legal traditions of Indigenous peoples in Canada. The section deconstructs the origins of Indigenous and non-Indigenous traditions by showing that while both are rooted in mythology, one prioritizes power and individuality, the other nature and community. The fixation of the non-Indigenous legal system on power and individuality is a tension point and the underlying reason for the subjugation of Indigenous legal systems.

Origin and Features of the Non-Indigenous (Western) Legal System

It is frequently the case that Indigenous legal traditions are subjugated to their non-Indigenous counterpart on the false assumption that they are inferior; however, both systems have similar origins in mythology.[12] Manley-Casimir affirmed that the non-Indigenous legal system is founded on European cultural values that prioritize the status of the nation-state.[13] This tradition is traceable to ecclesiastical laws and Roman traditions over many centuries during which the church exerted influence on the world order, the authority of church and state being supreme and unquestionable. In addition to ecclesiastical and Roman laws, English law was largely influenced by Greek jurisprudence.[14]

In Anglo-Saxon England "secular and ecclesiastical courts were not sharply separated, and the two jurisdictions were hardly distinguished."[15] The absence of any separation between state and church, and between secular and ecclesiastical laws, meant that religious traditions were imported into norms governing secular relationships. These hardly distinguishable traditions eventually trickled down to the advent of common law traditions, which began after the Norman conquest of 1066 when medieval kings began instituting a system of writs consisting of royal orders for regulating social interactions; these traditions were preoccupied with the notion of legitimacy.[16]

Legitimacy can only be obtained by adhering to the norms acceptable to the church or state; these norms were written codes that were generally accepted as authoritative and unquestionable, founded in the belief that what is written is true and not to be challenged.[17] Basically, this means that non-Indigenous legal systems have a preoccupation with rules and authoritative norms governing human interactions, or create such rules and norms where none exist. Attached to the system is also the idea of objectivity and universality, which originated from the English courts of equity.[18] The idea of the objectivity of the non-Indigenous legal system is construed in the notion that judges are required to eliminate their subjective opinion in dispensing justice, and that such a value be universally accepted.[19]

Manley-Casimir writes that the "non-Indigenous legal system there-fore is characterized by state-centrism in which state law is constructed as authoritative. Through written and coded laws, that state regulates the lives of its members through the ideals of objectivity and universal values."[20] Interestingly, the non-Indigenous legal system is entrenched in the language of the law, the origin of which is entrenched in the ecclesiasticism of the Latin and Greek languages.[21]

Societal order and cohesion are values essential to these languages. Gordon Christie, while discussing Canadian law, noted that "the domestic legal system as an institution is a social and historical construct, a structure built on words and meanings, designed to promote certain values in an ordering system."[22] One other feature identified as relating to the non-Indigenous legal system is the centrality of the claim that non-Indigenous law is devoid of "mythology, narrative and collective memories."[23]

Origin and Features of the Indigenous Legal System

Indigenous legal systems are built on stories transferred orally through many generations. Within these stories are the wisdom, logic, and reasoning that allow for an understating the political structures of different Indigenous nations, the rules governing social interactions, and the relationship between the physical elements of the earth.[24] Indigenous legal systems emerged from "a combination of wisdom gleaned from mythological time and thousands of years spent reflecting on the best ways to live."[25] The stories and their transmission hold an essential role in Indigenous nations and provide crucial understanding of the environment and the role of Indigenous people in preserving the earth. Behind every story is an explanation that, when critically analysed, reveals the rules governing a given subject,[26] and as such "Indigenous stories embed law, legal principles, and legal processes."[27]

Tuma Young, a professor of Mi'kmaq studies, explained at the hearings for the National Inquiry into Missing and Murdered Women and Girls that in cultural practices and language one can find the legal principles inherently guiding the affairs of a given Indigenous people: "our principles come from our stories, our ceremonies, our songs, our languages, and our dances ... and most of our legal principles are there."[28] The specific legal structure of each Indigenous nation varies, but they are generally built on the principles of responsibility, respect, reciprocity, interconnectedness, and interdependency.[29] Indigenous laws are living laws that draw from the norm that individual and collective rights are dependent on living organisms such as land, water, animals, spirits, the rain, rivers, and the moon, with the aim of promoting safety and justice.[30]

It is naive to assert that Indigenous nations were lawless and savage prior to European contact. Indigenous nations have long-established political and legal orders that outlined the structures of each society. Examples abound to support this fact. For instance, the Haudenosaunee Nation's governance system included consensus, veto powers, and representative decision making;[31] Plains Blackfoot and Cree law had no concept of private ownership of land as land is an object to be shared with all.[32] In many Indigenous nations, perpetrating violence on other clan members was not permitted and such acts were punished accordingly.[33]

In Indigenous nations, while collective rights are individualized, in the sense that they are exercised by individual members, priority is placed on the collective, which is contrary to the individualistic nature of the non-Indigenous legal system.[34] The priority of Indigenous law is the concept of responsibility – the responsibility to other members of the clan, to nature, to the environment and the spiritual world.[35] This is perfectly summed up in the following: "some people call it the Great Law, or the Great Law of Peace, and it is. This law, our law, does not define 'rights'; it does not defend 'rights.' In our ways, there are no 'rights,' only responsibilities: to observe the clans, to bring honour, trust, friendship and respect; to share; to be kind, honest and knowledgeable; to maintain a relationship with all the natural world."[36]

Privileging the Non-Indigenous Legal System

Some have argued that Indigenous peoples are pre-legal because they do not possess the positivist orientation and non-Indigenous conceptualization of power and authority.[37] The problem with this narrative is its intentional mischaracterization of Indigenous legal traditions; the mere fact that Indigenous institutions are not structured in the legalistic framework of non-Indigenous legal systems does not invalidate their authenticity and relevance. As John Borrows writes,

> In fact, despite the doubts some might hold concerning the presence of law in indigenous societies, there has been a long history of recognition of indigenous peoples' government or legal traditions by those who encountered these societies. Europeans' pronouncements that indigenous people had no government or law were contradicted by their practice of dealing with them through treaties and agreements. There was a long period of interaction between indigenous peoples prior to the arrival of Europeans and explorers from other continents. There were treaties, inter-marriages,

re-settlements, war and extended periods of peace. When Europeans and others came to North America, they encountered a complex socio-legal landscape. The complexity and scale of the interaction is demonstrated in early treaty and marriage relationships.[38]

Based on this classification of the Indigenous legal system as pre-legal and primitive, Indigenous people are excluded from non-Indigenous legal systems. The ideas of universality, neutrality, and the objectivity of European values were adopted as tools for the exclusion of Indigenous legal systems. However, universality is a myth conceived from the biased Eurocentric world view, and it privileges Eurocentric values of power, authority, and legitimacy.[39] It is a skewed and chauvinistic world view that paints Europeans as the carriers of civilization to the uninhabited and "lawless wilderness of North America."[40] The earth is generally amoral, and the origin of non-Indigenous legal system is not universal but culturally normative, thus the privileging of one tradition over another is immoral.[41]

Records of inter-communal bilateral relationships solidified through oral contracts detail the sophistication of Indigenous legal systems and contradict the false image of lawless and savage North America prior to colonialization. The Haudenosaunee and the Anishinabek had an oral agreement that was recorded on a wampum belt to the effect that the two nations would honourably share hunting grounds for food-gathering purposes.[42] This concept was subsequently adopted in the peace and friendship accord between the Mi'kmaq, Maliseet, Passamaquoddy, and the British Crown between 1685 and 1779.[43] Clearly, early interactions between colonial authorities and Indigenous nations were governed by Indigenous legal and cultural traditions.[44] Beyond treaty relations, European traders willingly agreed to Indigenous customs as governing laws regulating their business transactions, and marriages between European and Indigenous people were conducted under relevant customs.[45]

Laws are dynamic phenomena emerging from and evolving with cultural values and norms. Since norms are culturally relative, universality is impracticable and impossible. Therefore, to assert that North America was lawless prior to the arrival of Europeans is sanctimonious. The evolution of the Indigenous legal system is not tied to the promulgation of the *Indian Act*;[46] on the contrary, the introduction of the English common law was the genesis of the clash of two civilizations. Although recognized at the early stages, the promulgation of the *Indian Act* and the introduction of the English common law contributed to the suppression of Indigenous legal traditions.[47]

In 1867, the Quebec Superior Court in *Connolly v Woolwich*[48] recognized Cree law as forming part of the common law. As Justice Monk asked,

> Will it be contended that the territorial rights, political organisation such as it was, or the laws and usages of India tribes were abrogated – that they ceased exist when these two European nations began to trade with Aboriginal occupants? In my opinion it is beyond controversy that they did not – that so far from being abolished, they were left in full force, and were not even modified in the slightest degree.[49]

The judge in this case strongly believed that Indigenous legal traditions in existence prior to the arrival of English and French explorers should be and are part of the common law because the colonial relationship did not abrogate or modify Indigenous laws in the slightest degree. The assertion of Crown sovereignty was not intended to in any form terminate Indigenous laws because they "were presumed to survive the assertion of sovereignty and were absorbed into the common law as rights."[50]

Clash of Civilizations: Conflict of Laws? Or Crown Supremacy?

Despite the recognition of Indigenous law as forming part of the common law at the early stages of colonial interaction, subsequent rules were formulated limiting the application of Indigenous laws. These rules were to the effect that Indigenous laws were inapplicable in the following circumstances:[51] (a) if they were incompatible with the Crown's assertion of sovereignty; (b) if Indigenous nations by treaty voluntarily surrendered the applicability of Indigenous rules; and (c) where the government chose to extinguish them.

These rules practically infringed on Indigenous self-determination and autonomy – for Indigenous people, Crown supremacy, simply put, was an outright statement of subjugation. Given the reciprocal and respectful nature of the relationship during the first European contact, one wonders why the subjugation of customary laws began to take root as the colonial relationship evolved. Julie Evans, while discussing the use of international law as a tool of competition between imperial powers, noted that non-Indigenous laws were applied to "justify violence and discrimination against Indigenous peoples in order to dispossess them of their traditional lands."[52]

The subjugation of Indigenous laws would continue for many years until the institutionalization of Aboriginal rights in 1982 under section 35 of the *Constitution Act*.[53] Section 35 recognized the rights of Aboriginal peoples to self-determination, yet it's been suggested that the legal traditions of Indigenous nations were recognized by Canadian courts after 1982 only in those instances where doing so was deemed non-challenging to the Canadian nation-state.[54] For example, federal and provincial governments did not challenge the Spallumcheen Indian Band child welfare by-law outlining the band's exclusive jurisdiction over band children regardless of their place of residence;[55] Canadian law also recognized and adopted customary marriages and use of traditional processes for selecting band and council members.[56]

However, the state's recognition of Indigenous laws was possible because "these instances of jurisgenesis are seen as appropriate and non-threatening to the status quo and nation state's power."[57] The courts recognized non-threatening laws, yet they continued to reinforce the notion that recognition, appreciation, and acceptance of Indigenous laws was incompatible with maintaining state power.

Where Indigenous laws appear threatening, the Canadian legal system enters into battle mode by prioritizing its laws. A strong example is land rights and Aboriginal title. An indication of the priorities of the Canadian state with regards to lands is apparent from the fact that during land disputes, the state tends to violently exercise its "exclusive jurisdiction" through police brutality; the Oka stand-off and Ipperwash disputes are evidence of this tendency.[58] It also explains the terminology ascribed to Aboriginal title. Lindberg gives examples of the colonial impacts of Canadian legal language by pointing out that Aboriginal land rights are "claims" that must be proven and that the citizenship rights of Indigenous peoples are privileges that one must apply for.[59]

In many other ways, the supremacy of Canadian law and legal language was imposed upon Indigenous peoples. For example, the "point of contact" rule set by the Supreme Court of Canada in *Van der Peet*[60] sought to completely change the landscape of Aboriginal rights as guaranteed by section 35. In that case, two men with valid food fish licences caught fish near Chilliwack. Under the licence, they were legally permitted to fish for food purposes only. One of the men brought the fish to his partner, Dorothy Van der Peet, a Stó:lō woman from British Columbia, who then sold ten of the fish to a non-Indigenous woman. Van der Peet was charged with illegally selling fish under a licence intended only for food and ceremonial purposes. At trial, Van der Peet challenged the charges on the ground that section 35(1) of the *Constitution Act* protected her right to sell fish.[61]

Convicted by the magistrate, she mounted an appeal that went all the way to the Supreme Court of Canada (SCC). The SCC held that while the Stó:lō have the right to fish, this ancestral right does not include selling fish. The court established a new test for proving Aboriginal rights to wit that acts that constitute Indigenous rights must be of a distinctive cultural nature. Thus, to constitute an Aboriginal right, an activity must include an element of a custom, practice, or tradition forming an integral part of a distinct culture of the Aboriginal peoples.[62] The implication here was that Indigenous peoples cannot claim constitutional rights to customs and traditions that developed after European contact. By this very decision, the court stalled the growth of Indigenous jurisprudence on the ground of the "magical moment of contact" rule established in *Van der Peet*.[63] Indigenous peoples have inherent rights that are not dependent on them asserting and proving that their actions are distinctive and were in existence pre-European contact.

The SCC sadly followed this precedent in *R v Pamajewon*,[64] in which the Anishinaabe First Nations of Eagle Lake and Shawanaga argued in favour of the inherent rights of Indigenous peoples to self-government, particularly in respect of the right to control gambling practices on reserves and the rights to self-regulate communal economic activities. The SCC held that section 35 cannot protect the right to gambling because gambling was not proven to be a distinctive cultural practice of the two groups pre-contact. The court by this decision defined what constitutes Indigenous culture and insisted that only pre-contact activities qualify as Indigenous culture.[65] These decisions completely contradicted the ideal of self-determination that section 35 set out to achieve. Limiting people's rights and cultures to the moment of European contact implies that legal cultures are stagnant, and that Indigenous people are not permitted to evolve with the times and the unique circumstances in which they find themselves. It implies that Indigenous people are incapable of changing their social norms due to interactions with the non-Indigenous legal system or in response to basic survival needs.

As Manley-Casimir argues, "Indigenous communities within Canada continue to redefine their normative worlds in ways that support their continued cultural existence. Both forms of jurisgenesis – the revival of the old legal traditions and the development of the new legal traditions in the face of the colonial experience – are equally valid forms of legal interpretation."[66] Laws are not static – if they were, Canada ought to still be practising laws in existence at the time of Europeans' arrival to North America. Laws evolve with changes in circumstances, and to deny that Indigenous cultures must remain what they were at the point of first contact is an act of oppression. Similar to other legal systems, the Indigenous legal system has the ability to self-evolve, to develop from one reality to another without being forced in response to changing times and circumstances. Aboriginal rights are *sui generis* – inherently different and incomparable with any other – and the concepts on which such rights are based may not be accurately translatable to non-Indigenous languages or contexts.[67] Consequently, Aboriginal rights must be understood according to their own meanings and standards.

What, then, is the best framework for the implementation of Indigenous laws? Some have suggested incommensurability – the theory that two concepts are incomparable because there is no common standard for measuring them, and thus should not be compared. The coexistence of both systems is not strained by competition.[68] Applying this to the competition between Indigenous and non-Indigenous legal systems, it may be true that it is difficult to situate Indigenous legal concepts within the language of Canadian law, and that these legal systems are not easily translatable into non-Indigenous legal culture. However, Napoleon and Friedland believe that Indigenous legal traditions must be allowed to undergo critical analysis, like that applied to the non-Indigenous systems, and that the notion of incommensurability indicates fragility.[69] It has also been suggested that the Canadian legal system must adapt to and adopt a form of legal pluralism in which Indigenous laws can thrive.[70] An argument against legal pluralism is the idea that its suggests a competition between different legal orders in which one *nomos* must prevail.[71] To eliminate competition, Gordon Christie suggests a form of cooperative federalism that would enable fruitful relations between Canadian governments and Indigenous authorities.[72] While there is no general consensus on how exactly this would be implemented, there appears to be a general agreement that Indigenous legal system must be allowed to evolve and thrive in one form or another. The next section discusses Canadian law's prioritization of its own norms and the utilization of violence to stop the evolution of Indigenous law.

The Wet'suwet'en and the Coastal GasLink Pipeline Project

The Wet'suwet'en (People of the Wa Dzun Kwuh River) are the Indigenous people residing in an unceded territory in north-western British Columbia with an Athabaskan culture and relationship to the inland Dene groups.[73] Their territory is comprised of 22,000 square kilometres, which the people have lived on and used for thousands of generations.[74] They are a matrilineal society divided into clans, with each clan based on matrilineal kinship groups called *yikhs*.[75] Each clan exercises jurisdiction over some defined areas of the clan territory.[76] The right to the use of the land is considered collective in terms of access to resources and food, ceremonial uses, and economic pursuits generally; and ownership is viewed through the lens of responsibilities rather than rights.[77]

According to Daly,[78] prior to contact, the Wet'suwet'en people adopted migratory patterns in accordance with the season – they gathered food, supplies, and held ceremonies at villages and sites in the summer, after which they returned to their different *yikhs* for the winter season. The use of the lands by the clans is regularly validated through the feast of *baht'lat*, the Wet'suwet'en parliament and central governance institution.[79] The system of responsibilities is rooted in *yintahk*, meaning "everything is connected to the land," a cultural belief that the people do not just live on the land, they belong to it – they are inseparable from the land and everything connected with it.[80]

Governance and Legal Structures of the Wet'suwet'en

The Wet'suwet'en practised a decentralized system of government in which the *yikhs* congregate at the *baht'lat* to discuss collective decision on matters regarding the territory.[81] At the *baht'lat*, clan relationships and maintenance of the lands as well as inter-clan disputes are discussed and resolved.[82] Disputes over clan boundaries are typically presented at the *baht'lat* and negotiations over the same are conducted and resolved using traditional and oral histories, with reference to such natural features as creeks, rivers, or lakes.[83]

Appointment of clan hereditary chiefs are also validated during the *baht'lat* congregation, as are their rights and responsibilities.[84] Hereditary chiefs are selected while in the womb, by elder shamans and chiefs who may so decide after feeling an expectant mother's womb.[85] A child destined to be a chief is groomed from birth and instructed in the ways of the elders.[86] Succession rules include travelling into the bush to live and interact with the animals, learn their ways, and, upon returning, demonstrate the knowledge gained from the expedition.[87] At clan feasts and ceremonies, new chiefs are assigned their titles, robes, and crests, which confers on them the authority associated with their office.[88] The role of the hereditary chiefs is "to ensure the territory is managed in a responsible manner, so that the territory will always produce enough game, fish, berries and medicines to support the subsistence, trade, and customary needs of house members."[89]

According to the Office of the Wet'suwet'en, "the highest hereditary titles among the Wet'suwet'en are the twelve house chiefs. These twelve house chiefs own both fishing sites and distinct tracts of territory. The second highest titles or feast names are those of the twelve sub-chiefs who have important responsibilities for the administration of discrete parts of their House's territory."[90] This was the governance structure that thrived prior to European contact.

Elected Band Leadership System

Indigenous systems of governance came under renewed attack in 1869 when Canada promulgated *An Act for the Gradual Enfranchisement of Indians, the Better Management of Indian Affairs, and to Extend the Provisions of the Act 31st Victoria,* Chapter 42.[91] Section 10 of the act created the institution known as the band council to govern Indian affairs. This imposed European-style elections to undermine hereditary and other traditional leadership structures.

The *Indian Act* retained the band councils to manage reserves, but they held no jurisdiction over traditional territories, which continued to be under the leadership of hereditary chiefs.[92] However, the system disrupted the political governance structures of many Indigenous nations and disrespected the systems that had been in place for thousands of years.

Wet'suwet'en Governance: Hereditary Law versus Elected Band Councils

In 2012, LNG Canada selected TC Energy to build, own, and operate the Coastal GasLink project. After various additional negotiations, the construction project was publicly announced in October 2018.[93] According to the project's website, the Coastal GasLink project "will run approximately 670 km (416 miles) in length. The proposed pipeline will safely deliver natural gas from the Dawson Creek area of B.C. to a facility near Kitmat, B.C. where it will be converted to a liquid form by for export by LNG Canada."[94]

In a bid to evade the consultation process with the hereditary chiefs, Coastal GasLink exploited the leadership structure created by the non-Indigenous legal system – the elected band council of the Wet'suwet'en Nation. It is important to note that the elected band councils are funded by the federal government,[95] meaning that when confronted with a choice between tradition or economic imperatives, they may be tempted to defend policies that are traditionally unacceptable.

Coastal GasLink claimed that it signed project agreements with all twenty First Nations on the route of the pipeline, including the Wet'suwet'en Nation. The said agreements were signed by the elected band councils. The hereditary chiefs have asserted sovereignty over the unceded territory, as a result of which, the elected band council have no jurisdiction to enter into agreements that bind the Wet'suwet'en Nation. The Wet'suwet'en hereditary chiefs have asserted that their "free, prior and informed consent" was not obtained regarding the construction of the pipeline across their land.[96]

The hereditary chiefs asserted their authority over the territory on the strength of *Delgamuukw v British Columbia*.[97] The hereditary chiefs of the Gitxsan and Wet'suwet'en Nations had sought a declaration of Aboriginal title over 58,000 square kilometres in north-western British Columbia. At trial, the Province of British Columbia counterclaimed on the ground that the two nations had no right to or interest in the territory. The chiefs supported their "claim"[98] using oral evidence indicating use and possession of the lands in dispute, and they pointed to their spiritual connection to the land, including the feast hall, which identifies their connection with the area.[99] The trial judge rejected the oral evidence, opining that any title to the claimed territory was extinguished when British Columbia joined Confederation, and declaring the provincial government's right to unoccupied and vacant lands subject to the laws of the province.[100] The hereditary chiefs contested this decision up to the SCC.

The court held that "Aboriginal title is *sui generis*, and so distinguished from other proprietary interest, and characterised by several dimensions. It is inalienable and cannot be transferred, sold or surrendered to anyone other than the crown."[101] It also maintained that oral evidence is valid and must be treated as equal to other forms of evidence.[102]

Departing from the *Van der Peet* decision, the court held that it is sufficient to show that the occupied land was integral to the nations' cultures at the time of contact. To prove Aboriginal title, it must be shown that the claimants have maintained sufficient, continuous, and exclusive occupation of the lands. The court also held that traditional lands cannot be used in a manner that destroys the cultural value that was such a distinctive aspect the lands.[103]

The court also held that for government to infringe on Aboriginal title, it must meet an obligation to consult and provide fair compensation to the nations in question.[104] The court, however, ordered a retrial as a result of deficiencies relating to the pleadings.[105] Despite the fact that a new trial never occurred, the principles set out in the case are clear: title to unceded territories resides in the First Nation able to demonstrate its connection to the lands. The principles laid out in *Delgamuukw* were reaffirmed and restated in *Tsilhqot'in Nation v British Columbia*.[106]

The Wet'suwet'en hereditary chiefs relied on *Delgamuukw* to assert their sovereignty over the lands and denounce the Coastal GasLink project. They argued that the project would have adverse effects on the health, socio-economic viability, physical and cultural heritage, and current traditional uses of the lands by the Wet'suwet'en Nation.[107] They further asserted that, despite the SCC's pronouncement in *Delgamuukw*, government and private companies have continued to take steps on unceded territory that are "without good faith negotiation, treaties or agreement, consultation and accommodation, or free, prior, and informed consent."[108]

Acting on their sovereignty, the chiefs established three camps along the roads (including the Unist'ot'en Healing Centre and the cabin built on the exact location of the pipeline corridors)[109] in order to prevent pipeline workers from accessing the territory. These actions by the hereditary chiefs recall Manley-Casimir's observation that "Indigenous peoples within Canada continue to resist the violence of the non-Indigenous legal system. They continue to assert, despite assimilative governmental policies and laws, unique visions of their normative worlds through adherence to legal traditions that reflect their cultural values. Through such resistance, Indigenous peoples challenge the non-Indigenous legal system's clan to authoritative interpretation."[110]

The authoritative interpretation of the non-Indigenous legal system was confirmed by the RCMP's use of force in January 2019. Acting on an interim injunction obtained by Coastal GasLink in 2018, the RCMP invaded the Gidumt'en Camp and arrested fourteen clan members.[111] As justification for the invasion, the RCMP released a statement that included the following paragraph:

> For the land in question, where the Unist'ot'en camp is currently located near Houston, BC, it is our understanding that there has been no declaration of Aboriginal title in the Courts of Canada. In 1997, the Supreme Court of Canada issued an important decision, Delgamuukw v. British Columbia, that considered Aboriginal titles to Gitxsan and Wet'suwet'en traditional territories. The Supreme Court of Canada decided that a new trial was required to determine whether Aboriginal title had been claimed for these lands, and to hear from other Indigenous nations which have a stake in the territory claimed. The new trial has never been held, meaning that Aboriginal title to this land, and which Indigenous nation holds it, has not been determined.

This statement represents the interpretation that the non-Indigenous legal system chose to assign to *Delgamuukw* and aligns with that system's assertion of power and authority. It shows that Canadian institutions only understand the concept of power and authority and would go to any length to reinforce these notions. This fixation with power is also visible in the lack of cultural nuance and unfamiliarity with Indigenous legal tradition that judges often display in their decisions, such as the injunction granted against the hereditary chiefs by Justice Marguerite Church of the British Columbia Supreme Court.[112] The SCC decision in

Delgamuukw recognized that more than "mere consultation" is required in the context of unceded territory, yet the British Columbia Supreme Court granted an injunction despite the lack of consultation with Wet'suwet'en Indigenous governance structures.

Obviously, the Wet'suwet'en people are challenging the violence of the non-Indigenous legal system. Their struggle represents a vision for a future based on Indigenous legal systems. Although, the non-Indigenous legal system claims to acknowledge Indigenous legal frameworks, it continues to silence the vision and future of non-Indigenous legal systems. Nothing emphasizes this more than the Yellowhead Institute report that found that 76 per cent of injunctions filed by corporations against First Nations were successful, 82 per cent of injunctions by First Nations against federal and provincial governments were denied, and 81 per cent of injunctions filed against corporations by First Nations were also denied.[113] This suggests not only that the non-Indigenous legal system dominates and silences alternative legal forms by asserting state law, but that where the law recognizes Indigenous legal precedents as valid, it shackles such precedents to the dominant legal system's language or framework, thereby restricting the growth of the Indigenous legal system.

Canada's continued denial of the Indigenous capacity to evolve was clearly obvious in Justice Church's ruling in favour of Coastal GasLink's application for an interlocutory injunction: "The defendants may genuinely believe in their rights under Indigenous law to prevent the plaintiff from entering Dark House territory, but the law does not recognise any *right to blockade and obstruct* the plaintiff from *pursuing lawfully authorised activities.*"[114] The judge was quick to accept and describe the actions of Coastal GasLink as lawful and that of the First Nation as unlawful and not recognized in law. Part of the reasons adduced by the court related to competing claims between the hereditary chiefs and the elected band councils and Wet'suwet'en jurisprudence on who is entitled to make decisions on behalf of the people. The court admitted that the land was unceded and that, as such, negotiations should be held with the recognized Indigenous nations. That the court elected to align with the band council's declaration of the benefits of the project and disregard the hereditary chiefs' position on the danger of the project further confirms the priority of Canadian legal system.

The nuances that Justice Church failed to recognize and acknowledge related to the fact that the elected band council is a phenomenon that is unknown to Indigenous legal systems; it is thus not necessarily in alignment with the values inherent in the traditional governance structures of Indigenous nations. While it is possible that the evolution of

Indigenous law may have translated to something similar to a Western electoral system for leadership selection, the fact that the band councils are a creation and imposition of the Canadian legal system violates the sovereignty and self-determination of Indigenous peoples.

It may not be right to assert that band councils do not represent their nations' interests, but they are conflicted given their connection to the non-Indigenous legal system. Nothing in the *Indian Act* empowers band councils to make decisions in respect of unceded territory; rather, their jurisdiction appears limited to the reserves. The band council system is under the firm control of the government, which can disband, amalgamate, or constitute new bands[115] and declare testamentary documents of First Nations on reserves void.[116] The band council's conflict of interest was clearly outlined in *Coastal GasLink v Hudson*: "The elected Band councils assert that the reluctance of the Office of the Wet'suwet'en to enter into project agreements, out of concern that it might negatively impact their claims to Aboriginal title, placed the responsibility on the band councils to negotiate agreements."[117] Given this tension between the band councils and the hereditary chiefs, Justice Church, without questioning the grounds on which band councils can legally make decisions in respect of unceded territory, suggested that it is questionable whether hereditary protocol governance constitutes appropriate authority for decisions affecting the Wet'suwet'en Nation.[118] In granting the injunction, Justice Church appears to have suggested that the band council is an appropriate authority for making decisions in respect of unceded territory, in furtherance of the agenda of the non-Indigenous legal system.

Conclusion

Admittedly, Indigenous legal systems have suffered and continue to suffer from years of colonialism.[119] However, the renewed clashes with the non-Indigenous legal system will increase critical engagement and analysis of the strength and depth of Indigenous laws and traditions by both Indigenous and non-Indigenous legal scholars. It is essential to move Indigenous legal traditions from being viewed as "over-simplified pan-Indigenous explanations, often couched in terms of 'values' or 'worldviews,'" to being taken seriously as laws capable of being debated, analysed, and contested.[120] How can this be achieved? Training judges to understand Indigenous traditions may be insufficient because it is difficult for non-Indigenous judges to understand the nuances and to separate themselves from their inherent biases and their reliance on a legal tradition informed by their own history.

Lindberg suggests the need for the development of a collective critical consciousness of Indigenous laws. She believes that it is difficult to achieve a strong system so long as the English language remains the language of interpretation of Indigenous laws. Instead, a common language may need to be developed that "defines and situates our nations" and that "can strengthen the ways in which we address our critical existence."[121] It is unclear whether this means an actual language or some other communicative structure that allows for collective interpretation of Indigenous traditions. Napoleon and Friedland suggest using tools of legal analysis and synthesis that will allow for the building of the expertise required for developing and implementing Indigenous laws. They assert that Indigenous legal systems must be open to challenge so they can be more critically analysed like other legal traditions.

These are great approaches. However, while they are being tested or put into practice, Indigenous people must continue to use the tools of the non-Indigenous legal system to challenge mainstream legal frameworks and to promote the values of Indigenous legal traditions. The continued agitation by Indigenous nations such as the Wet'suwet'en brings these issues to national consciousness.

NOTES

1 Tracey Lindberg, "Critical Indigenous Legal Theory Part 1: The Dialogue Within," *Canadian Journal of Women and the Law* 27, no. 2 (2015): 224.

2 John Borrows, "Indigenous Legal Traditions in Canada," *Washington University Journal of Law and Policy* 19 (2005): 167.

3 Borrows, "Indigenous Legal Traditions in Canada," 167.

4 *Honouring the Truth, Reconciling for the Future: Summary of the Final Report of the Truth and Reconciliation Commission of Canada* (Winnipeg: Truth and Reconciliation Commission of Canada, 2015), 1, www.trc.ca/websites / trcinstitution/File/2015/Honouring_the_Truth_Reconciling_for_the _Future_July_23_2015.pdf. Hereafter cited as *TRC Report*.

5 *TRC Report*, 1.

6 *TRC Report*, 1.

7 Val Napoleon and Hadley Friedland, "An Inside Job: Engaging with Indigenous Legal Traditions through Stories," *McGill Law Journal* 61, no. 4 (2016): 725–54.

8 Napoleon and Friedland, "An Inside Job."

9 *TRC Report*.

10 John Borrows, *Recovering of Indigenous Law* (Toronto: University of Toronto Press, 2002), 125.

11 *TRC Report*, 205.

12 Kirsten Manley-Casimir, "Incommensurable Legal Cultures: Indigenous Legal Traditions and the Colonial Narrative," *Windsor Yearbook of Access to Justice* 30, no. 2 (2012): 137–61.

13 Manley-Casimir, "Incommensurable Legal Cultures."

14 Frederick Pollock and Frederic William Maitland, *The History of English Law: Before the Time of Edward I* (Cambridge: Cambridge University Press, 1895), 18.

15 Pollock and Maitland, *The History of English Law*, 18.

16 Pollock and Maitland, 18.

17 Manley-Casimir, "Incommensurable Legal Cultures."

18 Pollock and Maitland, *The History of English Law*.

19 Pollock and Maitland.

20 Manley-Casimir, "Incommensurable Legal Cultures," 140.

21 Manley-Casimir, 140.

22 Gordon Christie "Law, Theory and Aboriginal Peoples," *Indigenous Law Journal* 2 (2003): 69.

23 Manley-Casimir, "Incommensurable Legal Cultures," 141.

24 Napoleon and Friedland, "An Inside Job."

25 Christie, "Law Theory and Aboriginal Peoples," 91.

26 Napoleon and Friedland, "An Inside Job," 737.

27 Napoleon and Friedland, 737.

28 *Reclaiming Power and Place: The Final Report of the National Inquiry into Missing and Murdered Indigenous Women and Girls*, vol. 1a (Ottawa: National Inquiry into Missing and Murdered Indigenous Women and Girls, 2019), 136. Hereafter cited as *MMIWGI Report*.

29 *MMIWGI Report*, 136.

30 *MMIWGI Report*, 136.

31 Manley-Casimir, "Incommensurable Legal Cultures," 142.

32 Manley-Casimir, 142.

33 MMIWGI Report, 138.

34 Christie, "Law Theory and Aboriginal Peoples," 84.

35 Manley-Casimir, "Incommensurable Legal Cultures," 150.

36 Osennontion and Skonagenleh:rá, "Our World," *Canadian Woman Studies* 10, nos. 2–3 (1989): 11.

37 "One Tier Justice (Editorial)," *National Post*, 23 November 2004, A19, cited in Borrows, "Indigenous Legal Traditions in Canada," 176.

38 Borrows, "Indigenous Legal Traditions in Canada, 178.

39 Manley-Casimir, "Incommensurable Legal Cultures," 142.

40 Manley-Casimir, 142.

41 Manley-Casimir, 142.

42 Borrows, "Indigenous Legal Traditions in Canada, 179.

43 Borrows, 179.

44 Borrows, 179.

45 Borrows, 179.
46 *Indian Act*, RSC 1985, c I-5, amended by SC 2019, c. 25, c. 29.
47 *Indian Act*, RSC 1985, c I-5, amended by SC 2019, c. 25, c. 29.
48 Connolly v Woolrich, [1867] 17 RJRQ 75 (Can. QB. SC); Johnstone v Connelly, [1869], 17 RJRQ 266 (Can. QB QQB).
49 Connolly v Woolrich, [1867] 17 RJRQ 75 (Can. QB. SC); Borrows, "Indigenous Legal Traditions in Canada," 18.
50 R v Mitchell, [2001] SCR 911, 927; Borrows, "Indigenous Legal Traditions in Canada," 183.
51 R v Mitchell, [2001] SCR 911, 927; Borrows, "Indigenous Legal Traditions in Canada," 183.
52 Julie Evans, "Where Lawlessness Is Law: The Settler-Colonial Frontier as Legal Space of Violence," *Australian Feminist Law Journal* 30 (2009): 4.
53 *Constitution Act, 1982*, Schedule B to the *Canada Act, 1982*, c 11 (UK).
54 Manley-Casimir, "Incommensurable Legal Cultures."
55 Manley-Casimir.
56 Manley-Casimir.
57 Manley-Casimir, 143.
58 Manley-Casimir, 143.
59 Lindberg, "Critical Indigenous Legal Theory Part 1."
60 *R. v Van der Peet*, [1996] 2 SCR 507.
61 *R. v Van der Peet*, [1996] 2 SCR 507.
62 *R. v Van der Peet*, [1996] 2 SCR 507.
63 John Borrows "Revitalizing Canada's Indigenous Constitution: Two Challenges," in *UNDRIP Implementation: Braiding International, Domestic and Indigenous Laws*, ed. Brenda L. Gunn, Cheryl Knockwood, Gordon Christie, Jeffery G. Hewitt, John Borrows, Joshua Nichols, Lorena Sekwan Fontaine, Oonagh Fitzgerald, Risa Schwartz, and James (Sa'ke'j) Youngblood Henderson (Waterloo, ON: Centre for International Governance Innovation, 2017), 20.
64 *R. v Pamajewon*, [1996] 2 SCR 821.
65 *R. v Pamajewon*, [1996] 2 SCR 821.
66 Manley-Casimir, "Incommensurable Legal Cultures," 145.
67 John Borrows and Leonard Rotman, "The Sui Generis Nature of Aboriginal Rights: Does It Make a Difference?," *Alberta Law Review* 36, no. 1 (1997): 38.
68 Manley-Casimir, "Incommensurable Legal Cultures," 159.
69 Napoleon and Friedland, "An Inside Job."
70 Christie, "Law Theory and Aboriginal Peoples."
71 Geoffrey Swenson, "Legal Pluralism in Theory and Practice," *International Studies Review* 20, no. 3 (2018): 438.
72 Gordon Christie, "Indigenous Legal Orders, Canadian Law and UNDRIP," in Gunn et al., *UNDRIP Implementation*, 48.

73 Office of the Wet'suwet'en, *Wet'suwet'en Titles and Rights: Regarding Canada Department of Fisheries & Ocean and Pacific Trails Pipeline* (Smithers, BC: Office of the Wet'suwet'en, 2013), http://www.wetsuweten.com/files/PTP_FHCP _Response_to_DFO-25Nov13-Final.pdf.

74 Office of the Wet'suwet'en, *Wet'suwet'en Titles and Rights*.

75 Office of the Wet'suwet'en.

76 Office of the Wet'suwet'en, 7.

77 Office of the Wet'suwet'en, 7.

78 Richard Daly, *Our Box Was Full: An Ethnography for the Delgamuukw Plaintiffs* (Vancouver: UBC Press, 2005).

79 Office of the Wet'suwet'en, *Wet'suwet'en Titles and Rights*, 8.

80 Office of the Wet'suwet'en, 14.

81 Office of the Wet'suwet'en, 7.

82 Office of the Wet'suwet'en, 7.

83 Matthew Sparke, "A Map That Roared and an Original Atlas: Canada, Cartography, and the Narration of Nation," *Annals of the Association of American Geographers* 83, no. 3 (1998): 463–95.

84 Office of the Wet'suwet'en, *Wet'suwet'en Titles and Rights*, 8.

85 "Governance: Becoming a Hereditary Chief," Office of the Wet'suwet'en, accessed 27 November 2023, http://www.wetsuweten.com/culture /governance.

86 "Governance: Becoming a Hereditary Chief."

87 "Governance: Becoming a Hereditary Chief."

88 Office of the Wet'suwet'en, *Wet'suwet'en Titles and Rights*, 17.

89 Office of the Wet'suwet'en, 8.

90 "House Groups," Office of the Wet'suwet'en, accessed 27 November 2023, http://www.wetsuweten.com/culture/house-groups.

91 *An Act for the Gradual Enfranchisement of Indians, the Better Management of Indian Affairs, and to Extend the Provisions of the Act 31st Victoria, Chapter 42*, assented to 22 June 1869.

92 Delgamuukw v British Columbia, [1997] 3 SCR 1010.

93 "A Pipeline to Support the Liquefied Natural Gas (LNG) Industry," Coastal GasLink, accessed 15 February 2020, http://www.coastalgaslink.com /about/.

94 "A Pipeline to Support the Liquefied Natural Gas (LNG) Industry."

95 *Indian Act*, RSC 1985, c I-5; Anthony Gatensby, "The Legal Obligations of Band Councils: The Exclusion of Off-Reserve Members from Per-capita Distributions," *Indigenous Law Journal* 12, no. 1 (2004): 1–31.

96 Office of the Wet'suwet'en, *Wet'suwet'en Titles and Rights*, 88.

97 Delgamuukw v British Columbia, [1997] 3 SCR 1010.

98 The term "claim" is used loosely because the language of the non-Indigenous legal system changed the rights of Aboriginal people to claims that needed to be asserted and proven.

 99 Delgamuukw v British Columbia, [1997] 3 SCR 1010, para. 7.
100 Delgamuukw v British Columbia, [1997] 3 SCR 1010, para. 34.
101 Delgamuukw v British Columbia, [1997] 3 SCR 1010, Lamer C.J.
102 Delgamuukw v British Columbia, [1997] 3 SCR 1010, para. 85–7.
103 Delgamuukw v British Columbia, [1997] 3 SCR 1010, para. 128.
104 Delgamuukw v British Columbia, [1997] 3 SCR 1010, para. 168.
105 Delgamuukw v British Columbia, [1997] 3 SCR 1010, para. 208.
106 Tsilhqot'in Nation v British Columbia, [2014] 2 SCR 256.
107 Office of the Wet'suwet'en, *Wet'suwet'en Titles and Rights*, 89.
108 Office of the Wet'suwet'en, 89.
109 "Background of the Campaign," Unist'ot'en, accessed 27 November 2023, http://unistoten.camp/no-pipelines/background-of-the-campaign/.
110 Manley-Casimir, "Incommensurable Legal Cultures," 156.
111 Manley-Casimir, 156.
112 Coastal GasLink Pipeline Ltd v Hudson, [2019] BCSC 2264 (Can. BC SC).
113 Shiri Pasternak, Hayden King, and Riley Yesno, *Land Back: A Yellowhead Institute Red Paper* (Toronto: Yellowhead Institute, October 2019), https://redpaper.yellowheadinstitute.org/wp-content/uploads/2019/10/red-paper-report-final.pdf.
114 Coastal GasLink Pipeline Ltd v Hudson, [2019] BCSC 2264 (Can. BC SC), para. 225. Emphasis added.
115 *Indian Act*, RSC 1985, c I-5, s. 17.
116 *Indian Act*, RSC 1985, c I-5, s. 45.
117 Coastal GasLink Pipeline Ltd v Hudson, [2019] BCSC 2264 (Can. BC SC).
118 Coastal GasLink Pipeline Ltd v Hudson, [2019] BCSC 2264 (Can. BC SC), para. 134.
119 Napoleon and Friedland, "An Inside Job," 740.
120 Napoleon and Friedland, 740.
121 Lindberg, "Critical Indigenous Legal Theory Part 1," 230.

Chapter Twelve

Contemporary Colonialism: The Dakota Access Pipeline

HELENA ARBUCKLE

This chapter was written from my perspective as woman with Celtic (Scottish and Irish) and English ancestry, and as a settler who presently calls Ləkʷəŋən, Hupačasath, and čišaaʔatḥ lands home. I recognize my privileged experiences and opportunities at the expense of Indigenous lands, lives, world views, and livelihoods, and encourage fellow settlers to learn the long-standing story of genocide and dispossession that brings us to reside on forcibly taken lands.

The resilience and cultural and spiritual resurgence of the Oceti Sakowin at Sacred Stone Camp in the face of colonial violence moved me to write this essay. Where I reside in so-called British Columbia, Canada, similar camps have been established by the Secwepemc (Tiny House Warriors) and Wet'suwet'en (Unist'ot'en Camp) to block the Trans Mountain pipeline and the Coastal GasLink liquified natural gas pipeline, respectively. Through the resurgence and revitalization of Indigenous cultures, laws, and governance, Indigenous nations across Great Turtle Island are resisting colonialisms violence against Indigenous women, lands, and water. At the heart of these ceremonial movements is love for land, water, life, and all of the earth's inhabitants. This essay is dedicated to the Indigenous sheroes across Great Turtle Island who hold the line between what affirms and gives life, and what destroys, extracts, and takes without consent. Human beings with diverse heritages, nationalities, and cultures watch and follow your leadership with gratitude, solidarity, and support as you continue standing for life itself.

In loving memory of Genevieve Mary Gladue.

Tapadh leat Philámayaye Hiy, Hiy His'kwe ƛeekoo Uwala Thank You

Introduction

Mni Wiconi (water is life), or at least necessary for life; thus, to protect water is to protect life itself.[1] Across Great Turtle Island, there is a

Figure 12.1. "Solidarity with Standing Rock Sioux/No DAPL" action in front of TD Bank in Victoria, British Columbia.

Source: Courtesy of Colton Hash.

correlation between oppression and violence towards Indigenous people (particularly women) and the degradation of waterways and destruction of lands.[2] In other words, an inextricable connection exists in how we treat each other and how we treat the earth.[3] One example of many to support this claim is the United States of America's violent response to the peaceful action led by Oceti Sakowin (Seven Council Fires or Sioux)[4] women at Iŋyaŋ Wakháŋagapi Othí (Sacred Stone Camp) to stop the Dakota Access Pipeline (DAPL). Sacred Stone Camp was founded April 2016 by members of the Standing Rock Lakota Nation and allied Lakota, Nakota, and Dakota citizens (Oceti Sakowin) with the aim of stopping DAPL from crossing over the Missouri and Cannonball Rivers and contaminating the water. This essay argues that DAPL is exemplary of colonization as an ongoing process, system, and set of values, one that enables and perpetuates both violence against Indigenous women and violence against the earth.

The Oceti Sakowin are an ancient and sophisticated sovereign nation who have built institutions and systems of law, governance, economics, and stewardship over thousands of years, raising questions as to how jurisdiction over their homelands has been "lost."[5] To acquire someone else's homelands requires ongoing processes of violence, exclusion, assimilation, extraction, and replacement. These ongoing processes of colonialization are enacted by imposing oppressive and foreign legal, governance, economic, education, and familial systems shaped and informed by values of white, male, and human superiority. Colonization, as a process, system, and set of cultural values, is carried out by opaque, dishonest, predatory, and controlling actors. These actors are possessed by a superiority complex and sense of entitlement to violate the autonomy, self-determination, and consent of "others" with impunity. This essay explores the leadership of Oceti Sakowin women in resisting the disproportionate impacts of foreign processes (of violence, assimilation, and extraction), systems (of law, governance, and economics), and values (of patriarchy and anthropocentrism) imposed by colonial actors.

By design, violence against the land and water disproportionately harms Indigenous people.[6] For instance, the route selected for DAPL (crossing Fort Laramie Treaty land and running within half a mile of the Standing Rock Reservation) raises significant issues of racism, sexual violence, inequality, and white privilege. The route chosen for DAPL also exemplifies power relationships among and between actors in various systems of colonization. According to the Standing Rock Sioux Tribe's Motion for Preliminary Injunction, two routes for DAPL were initially considered.[7] Energy Transfer Partners rejected the northern route because it could endanger the drinking water of white settler Bismarck residents.[8] Instead, the oil company selected the southern route close to the border of the Standing Rock Reservation. This route crosses the confluence of the Cannonball and Missouri Rivers, threatening the drinking water of ten million people and destroying sacred burial grounds, historic villages, and sun dance sites surrounding the area.[9]

The Oceti Sakowin are subsequently bearing the costs and risks associated with DAPL without any foreseeable benefit. Moreover, there are no benefits from DAPL that could justify threatening the water and, consequently, every life dependent upon it. This is expressed in the following statement made by Dave Archambault, former chairman of the Standing Rock Reservation: "Our people will receive no benefits from this pipeline, yet we are paying the ultimate price for it with our water."[10] As such, Archambault argues that Indigenous nations in the area "will not stop asking the federal government and Army Corps to end their attacks on our water and our people."[11]

The Colonial Context of the Dakota Access Pipeline

Those who are fundamentally unjust in violence and war contravene the terms of an agreement or relationship, or violate the autonomy, self-determination, or consent of another. The United States waged a series of unjust wars against Indigenous nations after declaring independence from Britain in 1776, invading territories to exploit resources and privatize lands for white settlement.[12] Europeans violated and threatened the territorial integrity of the Oceti Sakowin's (Oglala Lakota) hunting land in so-called Powder River Country, starting the two-year Red Cloud War in 1866 throughout so-called Montana and Wyoming.[13] The Oglala Lakota leader Makhpyia-luta (Red Cloud) successfully waged war against the United States, resulting in the 1868 Treaty of Fort Laramie.[14]

Treaties are constitutionally significant, legally binding agreements that set out responsibilities and relationships between nations. According to US constitutional law, treaties are the supreme law of the land.[15] Europeans were compelled to negotiate treaties because Indigenous nations were here first, and these agreements are the only way for settler governments to claim Indigenous lands legitimately under British common law.[16] The Fort Laramie Treaty was signed between Oceti Sakowin leaders and the US government, guaranteeing Oceti Sakowin jurisdiction over twenty-six million acres of their land, Paha Sapa (the sacred Black Hills), and establishing the Great Sioux Reservation.[17] The Fort Laramie Treaty has four key parts. First, it pledges peace between both sides.[18] Second, the treaty maintains that if "bad men among the whites … commit any wrong upon the person or property of the Indians, the United States will … proceed at once to cause the offender to be arrested and punished."[19] Third, it claims that Oceti Sakowin are to have unrestricted use of their territory east of the Rockies and west of the Missouri River.[20] And fourth, it states that the US government cannot change the terms of the treaty without approval from three-quarters of adult Oceti Sakowin *men*.[21]

Despite the fact that the terms of the Fort Laramie Treaty cannot be changed without Oceti Sakowin approval, in 1874 (just eight years after the treaty's ratification) the US government annexed 7.7 million acres of Oceti Sakowin land (including Paha Sapa) and water (including rivers, lakes, streams, and reservoirs).[22] The United States attempted to negotiate another treaty to cede and purchase Paha Sapa rather than uphold its treaty responsibility to stop the incursion of settlers, such as George Custer's military expedition through Paha Sapa to extract minerals.[23] When these new treaty negotiations failed, the United States launched another unjust war against the Oceti Sakowin and Cheyenne lasting from 1876 to 1877, the Great Sioux War.[24] Then, in 1877, the US Congress passed a statute that unlawfully ignored the terms of the Fort Laramie Treaty and stole Oceti Sakowin homelands, waterways, and, as a result, livelihoods.[25]

The US legal system continues to allow the government to exercise congressional power and ignore its treaty obligations despite their constitutional recognition. For instance, the unlawful confiscation of Oceti Sakowin land was not subject to litigation because of US "sovereign immunity."[26] Litigation concerning the illegal seizure of Paha Sapa finally began in 1923 after Congress passed a special jurisdictional statute waiving sovereign immunity. In 1980, the Supreme Court finally determined that the United States was "exercising its power of eminent domain under the Fifth Amendment" when it "took" Paha Sapa, and that the US government's only wrongdoing was failing to promote "just compensation."[27] The Supreme Court dismissed the fact that the US government contravened agreements made under the Fort Laramie Treaty[28] and violated the autonomy, self-determination, and consent of the Oceti Sakowin. The US government continues to ignore and violate the terms of the Fort Laramie Treaty through its use of military power, and it forces the Oceti Sakowin to participate within a foreign, imposed, and colonial legal system.

Due to the Supreme Court's consistent failure to uphold Oceti Sakowin treaty rights in the Black Hills (Paha Sapa) as part of its judicial review,[29] it is unsurprising that the Sioux Tribe did not claim a violation of the Fort Laramie Treaty when filing an injunction to stop DAPL in July 2016.[30] DAPL (owned and operated by Energy Transfer Partners) is a domestic oil pipeline route moving over half a billion gallons of crude oil across four states daily.[31] There is an ancient Lakota prophecy about a black snake crossing the land, desecrating sacred sites, poisoning the water, and destroying the earth. To many, DAPL is that black snake. As Dave Archambault notes, "There was a prophecy saying that there is a black snake above ground. And what do we see? We see black highways across the nation … There's also a prophecy that when that black snake goes underground, it's going to be devastating to the Earth."[32] The prophecy also notes that the Oceti Sakowin were to come together to defeat the black snake.[33] Neither the US government, the US Army Corps of Engineers, nor Energy Transfer Partners obtained consent from the Sioux Tribe for the pipeline route, leaving the Oceti Sakowin no choice but to defend themselves and protect the water against the black snake.[34]

Systemic limitations and oppression are what fuel front-line resistance, direct action, and stand-offs between police and peaceful land and water protectors. Over 280 Indigenous nations from around the world joined the Oceti Sakowin's opposition to DAPL,[35] with thousands of Indigenous and non-Indigenous allies travelling to Standing Rock as well as solidarity actions occurring globally. At the front line of the resistance to DAPL, the terms of the Fort Laramie Treaty continued to be violated numerous times. For example, police used excessive and almost deadly force against water protectors on 21 November 2016.[36] Over 180 people were injured by water cannons in sub-zero temperatures, rubber bullets, chemical irritants, and concussion grenades.[37] Moreover, the Obama administration took no action to stop the violent and forceful removal of Oceti Sakowin water protectors from their homelands.[38]

Leadership, Resistance, and Revolution from the Front Lines

Despite the forceful removal of protectors from their homelands and the use of police violence, Oceti Sakowin women revitalized matriarchal tribal structures and assumed their role as spiritual leaders.[39] From the front lines, Jasilea Rose Charger from Cheyenne River Sioux Tribe stated that resistance to DAPL is not a "war of negativity but the war of spirituality. We're trying to overcome this obstacle in a non-violent way. Instead of fighting with violence, instead of fighting with fire, we're fighting with prayer."[40] Moreover, Oceti Sakowin woman Kandi Mossett asserted, "I am protecting the very essence of what I am made up of, which is mostly water ... Water is the first life – it is our very essence, our very being is made up of water."[41] The opposition to DAPL is therefore a demonstration of profound leadership aimed at protecting life itself, and the label "protestor" is too narrow to describe the purpose, power, and impact of those who gathered at Sacred Stone Camp.[42]

Assimilation, as an ongoing process of colonialism, necessitates the creation of power imbalances between Indigenous men and women.[43] Recall, for instance, that changing the terms of the Fort Laramie Treaty required approval from three-quarters of adult Oceti Sakowin men and not women. Colonial governments refused to recognize and respect the decision-making and political authority of Indigenous women in intergovernmental relations, including in treaties. This, coupled with the imposition of foreign, patriarchal systems of law, governance, and economics, disenfranchised Indigenous women from leadership roles and decision-making processes, and even from membership in their own nations. As Makere Stewart-Harawira aptly writes, "As Christianity and capitalism spread throughout the world, recognition of the sacred and political roles of Indigenous women was one of the greatest casualties. Yet many Indigenous women have continued to exercise significant political and spiritual leadership."[44] For many Oceti Sakowin water protectors and spiritual leaders, the movement at Sacred Stone Camp is therefore an ongoing fight against misogyny, racism, and violence.[45]

Both spatially (across various settler-colonial states) and temporally (within the long-standing history and legacy of colonization) there is a correlation between violence against Indigenous women and violence against the land and water. To protect and care for water, the lifeblood of Mother Earth, is to protect and care for all of the earth's inhabitants (including ourselves). Similarly, to protect and care for women, the life givers of our species, is to ensure the health, well-being, and continuation of our families, communities, and nations. However, do women and water need protection? Or, alternatively, do we need to be respected, acknowledged, and adhered to as autonomous, self-determining beings with meaning, purpose, and value within this interconnected and interdependent web of life and relationships?

In order to stop the theft and murder of Indigenous women and girls, and the consequent genocide of Indigenous people, women-lead resistance and cultural resurgence camps across Great Turtle Island are calling for the abolition of "man camps."[46] In North Dakota, there are large numbers of non-Native men relocating to man camps or shantytowns near reservations. Since the 2008 oil boom and the subsequent increase in the number of man camps, North Dakota has seen more crime and violence against Indigenous people, especially women.[47] Similarly, in so-called Canada, *Reclaiming Power and Place: The Final Report of the National Inquiry into Missing and Murdered Indigenous Women and Girls* dedicates an entire section to this correlative increase in violence. Furthermore, Muscogee (Creek) lawyer Sarah Deer maintains that "the crime that Native women are experiencing as a result of the exploding fracking business has parallels with the harm being done to the planet – the land and water are being poisoned as the hearts and spirits of Native women break. Thus another generation [of Indigenous women] experiences displacement and abuse."[48] The displacement and abuse brought by man camps is a continuation of frontier dynamics outlined in the colonists' histories of discovering gold, oil, and other minerals in the region.[49] As a response, Oceti Sakowin women continue to resist the violence perpetuated against their bodies, minds, spirits, lands, and waters.[50]

In *Honouring Our Relations: An Anishinaabe Perspective on Environmental Justice*, Deborah McGregor articulates an Indigenous approach to environmental justice predicated on the necessity of achieving justice for all the Creator's beings.[51] Oceti Sakowin leader Kandi Mossett upholds this understanding of justice and natural law with the following statement: "I am protecting [the water] for my future generations ... all those that can't speak for themselves – not just the babies, but everything that flies in the sky, all those that swim in the waters, [and] the four-leggeds. Somebody has to speak on their behalf because they don't have a voice."[52] As McGregor maintains, both humans and non-humans have roles and responsibilities to fulfil that must be respected.[53] Thus, McGregor argues, "We do not have the right to interfere with water's duties to the rest of Creation."[54]

Indigenous epistemologies value water as the lifeblood of Mother Earth.[55] As such, McGregor maintains that, "just as water is a giver of life, women, also life givers, have a special relationship and responsibility to water. The recognition of women's role in creating life along with water means that women and water have a special bond."[56] Mossett speaks to this bond in the following statement: "It is no coincidence that when we're pregnant we carry our babies in water, and the understanding is that water is the first life – it's our very essence, our very being is made up of water. It flows through us, and it flows from the rivers to the sky [and] back down in that circular way."[57] To continue the cycle and natural order of things therefore requires an appreciation and respect for water as a living and life-giving entity.[58] As such, although Indigenous cultures are unique and diverse, most nations consider Indigenous women as the keepers of the water.

In leading the movement to stop DAPL, Oceti Sakowin women are therefore exercising their responsibility to protect the water. Simultaneously, they are raising awareness of the spiritual and cultural significance of water.[59] As Kandi Mossett notes, "Women are the ones going and breaking down fences and running in front of bulldozers. Women are the ones locking arms with babies on their backs going in because it's that desperate, it's that urgent to protect life."[60] Exemplary of this urgency, young women leading the Standing Rock Youth Council have faced physical violence in the form of mace, tear gas, and rubber bullets during increasingly tense stand-offs with police.[61] Despite the violence and the constant fear, threats, and arrests, Oceti Sakowin women elders held prayer circles directly on land where DAPL construction was planned.[62] Thus, as Mossett aptly claims from the front lines of the DAPL resistance, "We're all, as women, going to keep on holding that line pushing forward, and hold the line for water and for life."[63]

Conclusion

Due to Oceti Sakowin women holding the front lines of DAPL resistance, the Obama administration declined the final permit for the pipeline's construction in December 2016. However, in January 2017, then President Donald Trump signed an executive memorandum calling for the expedited approval of the DAPL and Keystone XL pipelines, in addition to an executive order on pipeline construction that shortens environmental review processes and streamlines regulations.[64] Trump also wanted his face carved onto the sacred Paha Sapa alongside the other four colonial actors that scar so-called Mount Rushmore – a national monument to settler colonialism and white male superiority.

As this essay argues, DAPL exemplifies colonization as an ongoing process, system, and set of values that enables and perpetuates both violence against Indigenous women and violence against the land and water. It is therefore disturbing, yet ultimately unsurprising, that an entitled, dishonest, predatory, and controlling "man" with a superiority complex would violate the autonomy and self-determination of the Sioux Tribe. Trump was, after all, elected and subsequently treated with impunity despite multiple conflicts of interest, being accused of sexual assault, and blatantly bragging about violating women. Although this particular battle at Standing Rock may not have been won, the 245-year-old war between the United States and Oceti Sakowin continues – this time, with more solidarity and allyship than ever before as Indigenous women remind us of the necessity to hold the line for water, land, and life itself, and how to be good leaders, healers, stewards, and human beings.

NOTES

1 Patrick Wolfe, "Settler Colonialism and the Elimination of the Native," *Journal of Genocidal Research* 8, no. 4 (2006): 387, https://doi.org/10.1080/14623520601056240.
2 Gretta Gaard, "Women, Water, Energy: An Ecofeminist Approach," *Organization & Environment* 14, no. 2 (2001): 159.
3 Gaard, "Women, Water, Energy," 158.
4 "Oceti Sakowin," Akata Lakota Museum Cultural Center, accessed 3 October 2024, https://aktalakota.stjo.org/oceti-sakowin-seven-council-fires/. Seven major "tribal" nations associated with the Indigenous peoples commonly known as "Sioux" comprise the Oceti Sakowin.
5 Michael Asch, *On Being Here to Stay: Treaties and Aboriginal Rights in Canada* (Toronto: University of Toronto Press, 2014), 3.
6 Deborah McGregor, "Honouring Our Relations: An Anishinaabe Perspective on Environmental Justice," in *Speaking for Ourselves: Environmental Justice in Canada*, ed. Julian Agyeman, Peter Cole, Randolph Haluza-Delay, and Pat O'Riley (Vancouver: UBC Press, 2009), 27.
7 Standing Rock Sioux Tribe et al. v US Army Corps of Engineers, 16–1534 (DC District Court, 2016).

8 Standing Rock Sioux Tribe et al. v US Army Corps of Engineers, 16–1534 (DC District Court, 2016).

9 Iyuskin American Horse, "We Are Protectors, Not Protesters: Why I'm Fighting the North Dakota Pipeline," *The Guardian*, 18 August 2016, https://www.theguardian.com/us-news/2016/aug/18/northdakota -pipeline-activists-bakken-oil-fields.

10 Valerie Taliman, "Dakota Access Pipeline Standoff: Mni Wiconi, Water Is Life," *ICT News*, 16 August 2016, https://ictnews.org/archive/dakota -access-pipeline-standoff-mni-wiconi-water-is-life.

11 Taliman, "Dakota Access Pipeline Standoff."

12 Gabriel S. Estrada, "Battle of Little Bighorn," in *Multicultural America: A Multimedia Encyclopedia*, ed. Carlos E. Cortés (Thousand Oaks, CA: SAGE Publications, 2013), 2, http://dx.doi.org/10.4135/9781452276274.n5.

13 Estrada, "Battle of Little Bighorn."

14 Estrada, 2.

15 Frank Pommersheim, *Broken Landscape: Indians, Indian Tribes, and the Constitution* (Oxford: Oxford University Press, 2009), 4.

16 Pommersheim, *Broken Landscape*, 2.

17 Lisa Roberts Seppi, "Fort Laramie, Treaty of (1868)," in Cortés, *Multicultural America*, 2.

18 Fort Laramie Treaty, Sioux-Brulé-Oglala-Minconjou-Yanktonai-Hunkpapa-Blackfeet-Cuthead-Two Kettle-Sans Arcs-Santee-Arapaho, 29 April 1868, art. 1, https://standingrock.org/about/history/fort-laramie-treaty/.

19 Fort Laramie Treaty, art. 1.

20 Fort Laramie Treaty, art. 2.

21 Fort Laramie Treaty, art. 12.

22 Pommersheim, *Broken Landscape*, 75.

23 Seppi, "Fort Laramie, Treaty of (1868)," 2.

24 Seppi, 2.

25 Pommersheim, *Broken Landscape*, 75.

26 Pommersheim, 75.

27 Pommersheim, 75.

28 Pommersheim, 75.

29 Pommersheim, 75.

30 Standing Rock Sioux Tribe et al. v US Army Corps of Engineers, 16–1534 (DC District Court, 2016). The Sioux Tribe argued that upon approving DAPL, the US Army Corps of Engineers violated multiple statutes under US federal law, including the *Clean Water Act*. Further, the Sioux Tribe argued in court that DAPL is an "unlawful encroachment on its [the Sioux Nation's] heritage."

31 Standing Rock Sioux Tribe et al. v US Army Corps of Engineers, 16–1534 (DC District Court, 2016).

32 Karen Pauls, "'We Must Kill the Black Snake': Prophecy and Prayer Motivate Standing Rock Movement," *CBC News*, 11 December 2016, https://www.cbc.ca/news/canada/manitoba/dakota-access-pipeline-prayer-1.3887441.

33 Sacred Stone Camp, "Standing Strong," Facebook video, 19 September 2016, https://www.facebook.com/CampOfTheSacredStone/videos/1773941612894869/.

34 Taliman, "Dakota Access Pipeline Standoff."

35 Sacred Stone Camp, "Standing Strong."

36 Associated Press, "Standing Rock Pipeline Protesters Repelled by Force at Bridge Crossing," *CBC News*, 21 November 2016, http://www.cbc.ca/news/world/dakota-access-clashes-sunday-night-1.3859945.

37 Associated Press, "Standing Rock Pipeline Protesters Repelled."

38 Associated Press.

39 Sam Levin, "At Standing Rock, Women Lead Fight in Face of Mace, Arrests and Strip Searches," *The Guardian*, 4 November 2016, https://www.theguardian.com/us-news/2016/nov/04/dakota-access-pipeline-protest-standing-rock-women-police-abuse.

40 Sacred Stone Camp, "Standing Strong."

41 Fusion, "Dakota Access Pipeline: Protectors Not Protesters," YouTube, 8 September 2016, accessed 13 October 2017, https://www.youtube.com/watch?v=U8Uwo6ZAEG4.

42 Fusion, "Dakota Access Pipeline."

43 Denise K. Henning, "Yes My Daughters, We Are Cherokee Women," in *Making Space for Indigenous Feminism*, ed. Joyce Green (Winnipeg: Fernwood Publishing, 2007), 189.

44 Makere Stewart-Harawira, "Practicing Indigenous Feminism: Resistance to Imperialism," in Green, *Making Space for Indigenous Feminism*, 133.

45 Levin, "At Standing Rock, Women Lead Fight."

46 Sarah Deer, *The Beginning and End of Rape: Confronting Sexual Violence in Native America* (Minneapolis: University of Minnesota Press, 2015), 78.

47 Deer, *The Beginning and End of Rape*, 78.

48 Deer, 78.

49 Deer, 78.

50 Deer, 78.

51 McGregor, "Honouring Our Relations," 27.

52 Fusion, "Dakota Access Pipeline."

53 McGregor, "Honouring Our Relations," 32–3.

54 McGregor, 38.

55 McGregor, 38.

56 McGregor, 38.

57 Fusion, "Dakota Access Pipeline."

58 Fusion, "Dakota Access Pipeline."

59 McGregor, "Honouring Our Relations," 39.
60 McGregor, 39.
61 Levin, "At Standing Rock, Women Lead Fight."
62 Levin.
63 Fusion, "Dakota Access Pipeline."
64 Athena Jones, Jeremy Diamond, and Gregory Krieg, "Trump Advances Controversial Oil Pipelines with Executive Action," *CNN Politics*, 24 January 2017, https://www.cnn.com/2017/01/24/politics/trump-keystone-xl -dakota-access-pipelines-executive-actions/index.html.

A RAVEN Roundtable

JOHN BORROWS, GLEN COULTHARD, MIKE FABRIS, DAWN HOOGEVEEN, MAX RITTS, AND SUSAN SMITTEN

The following is an edited transcript from conversations that took place over the summer of 2021 and the summer of 2022. The speakers, in order of appearance, are Susan Smitten (SS), John Borrows (JB), Glen Coulthard (GC), Mike Fabris (MF), and Max Ritts (MR). The conversation was edited by Max Ritts, Dawn Hoogeveen, and Maëve Leduc.

SS: Hi everyone. Thank you all so much for joining on this roundtable! To start, what made you willing to get involved in the adjudication of the RAVEN essays in the first place?

JB: I got involved in this project because I wanted to learn from the students. Knowing RAVEN's work, through the years as an adviser, and understanding the link between Indigenous issues and environmental issues, I was interested in seeing how students engaged in this work.

GC: I've been involved for probably five years, so I'm kind of grandmothered in. But I agree with John's assessment, it's about the opportunities for students.

MF: I thought it was great to have a project that supports students who are exploring these intersections of Indigenous and environmental issues. It's not always a given, especially with respect to how environmental issues in Canada and the United States have been written about or organized around by the mainstream environmental movement.

Also, it's clear from looking at the papers over the years that people are writing about things that really are important to them. In a lot of cases, they are directly invested in them personally, and that's not always the kind of work that gets rewarded – especially later within different stages of people's degrees, or academic careers. Depending on the discipline it could be different, but I think in a lot of cases this is not the kind of work that academia rewards or incentivizes. So having a reward for this kind of work, I also was inspired by that.

MR: Thanks for this. When each of you began your careers as academics, there was perhaps *some* opportunity to bring Indigenous perspectives to bear on environmental issues, but certainly many more limitations on what the academy was willing to accept. I'm curious to hear your thoughts on what changed over the last few decades, or decade, since each of you have moved from being students to teachers?

JB: I've been teaching almost thirty years now, and many things have changed. When I did my graduate work, the place to start research was in the community. Once I talked to my elders, chief, and family, *then* I went to the archives to piece together what I heard, such as documents about the Treaty of Niagara, the War of 1812, the introduction of the *Indian Act*, and other challenges throughout the generations.

I was encouraged by a new professor in his second year of teaching, named Patrick Macklem, who supervised my LLM. Ken McNeil was my PhD supervisor; he was writing in my field and had been at the Native Law Centre before going to Osgoode Hall Law School. There weren't many Indigenous people in the academy, so I chose to work with professors who respected our communities. Their support allowed me to search for answers to my questions outside of the academy.

GC: I've been teaching now for twelve or thirteen years. I've had the benefit of having people like John in the academy that I can reference and speak with, so it's a bit different. I tried to push back I guess on a puritanical, at some points, perspective on Indigenous studies and Indigenous thought as being wholly traditional or what have you. I hope to have developed through an engagement with my own community's history, which is the Dene, and around the pipeline and treaty issues, a perspective that is still a Dene perspective, but a perspective that deals with other traditions; Third World theorists like Frantz Fanon, or Karl Marx, but from a Dene perspective that's grounded in the land. I think that was only possible because of people like John [Borrows], who paved the way to make that Indigenous perspective legible in the academy.

MF: I started my undergrad in 1997. There is no question things were different than they are now. There was a huge sense of isolation, and a sense that anything that I was doing related to Indigenous people or my community was just really off the radar. I remember I was involved in the First Nations Student Association at SFU. We had no funding; we had a room. Basically, the services for students were a place to hang out and kvetch [*laughter*]. But also to approach real experiences of racism and isolation.

Fast-forwarding to 2014, when I started my master's, it was a completely different space. When I made the decision to return to university, I still expected that I was going to be relatively isolated. Glen was one of the first people I talked to about bringing in the things I've been thinking about for years as an activist to a more academic setting. That sort of mentorship for people has been really important. And also to learn from Indigenous scholars like Dr. Sarah Hunt and Dr. Michelle Daigle.

ss: Are you sensing a transformation then? Are you sensing that there's a shift not just in what's being taught, but *why* it's being taught?

mf: I think so. A couple years ago now, or a year and a half, by the beginning of the pandemic I was teaching a public policy class at SFU, and I was really going in expecting to have to deal with a lot of pushback from what I was teaching. I even had a bit of a warning at the beginning of the class – you know, that if you want to come in here and play devil's advocate and treat these collective experiences of trauma like an opportunity to win a debate, this is not the class for you. And nobody left, nobody got their backs up about that kind of approach. I'm sure it does happen – I know that it does – but it was different than what I was expecting.

gc: The access to knowledge, and bodies too, is just more diverse today. And that's always a good thing. But I don't know how much.

jb: Students today can engage in a way that's more alive to comparison and contrast.

There is a saying, that a fish doesn't know about water until the first time that it jumps into the air. Often, we swim around in the academy without really understanding the choices behind what we're learning. When introducing Indigenous law or Indigenous perspectives on law or governance, students are thrown out of their element, and they have to deal with contrast; they recognize that there's something distinct with Indigenous issues. This can help them question what they're learning in their other courses.

So, to get more concrete, the students are drawing in insights from Western views of land-use planning, governance, philosophy, or social sciences, along with comparisons with Indigenous perspectives and knowledge at the same time. Contrast deepens our understanding of Indigenous fields, and also challenges Western perspectives on land-use planning, sociology, or law, or whatever the discipline might be. And I really appreciate this in the students' essays in this book.

ss: Which sort of leads to the next question around this idea of redefining the framework of university teaching in particular. There is a need to think institutionally, not just by adding Indigenous content, but by achieving full-on transformation, which requires this deeper shift in thinking. To what extent do you see signs of a shift to either Indigenize or actually decolonize university education?

gc: I waver between being sceptical and positive on that. With Dechinta Centre for Research and Learning, it was quite an uphill battle to get the accreditation system in place. It's been a challenge getting them [university administrators] to understand the kind of practices that we engage in as facilitating critical thinking and so on. There are some institutional constraints, I think, built into universities that you have to contend with. But when you go to the community to teach and try to establish some sort of autonomy over curriculum development and evaluation and so on, you can really have the effect of having an autonomous-type institution have influence on the more colonial legacies in universities.

JB: Like Glen, I think the answer is it depends. In some places, you see universities change because of the work of Indigenous scholars, students, and communities. This occurs with the support of allies, and it helps to change curriculum, programming, who's hired, and how we make decisions about those things. Like Glen, I've also really appreciated community-based learning as a place to think about transformation.

SS: I'm curious from your perspectives how it might appear in these student essays?

MF: Well, I wouldn't say it applies to every paper, but it's more the norm for the papers to have a whole practice of the author situating themselves within their work. And that is something to me that's within both Indigenous and feminist research methods. It's been a tradition that I think people have been pushing for for a long time. But it's still not the norm; reading the student papers versus recently peer-reviewed journals, it's clear from what I'm reading that even research that is for or on Indigenous peoples, like in geography journals, that practice of self-location is still not the norm, and there still can be outright hostility for it when people bring up those questions – you know, as a reader asking, "Where do you situate yourself within the work?"

JB: I also really liked the self-location of the authors in this book, identifying the place from which they're writing, which in some cases led to narrative approaches in their work.

MR: It was interesting to hear Glen reflect on how John was a big part of his own kind of coming into the academy. It's a simple reminder of the importance of having someone who could provide an example of how to find one's voice in a not always friendly academy setting.

MF: It's a huge change, and for me it's an important thing. It's not just the type of research, but there is a skill set for surviving within the university, and pushing back within the university. And also, that relatability, that when we enter the university, the lives that we have still exist, there's still everything else that's happening outside of the university that comes along with being Indigenous in Canada. And I think the way that fits into this is enacting areas of empathy or collective care. To me that's been key to being able to stay within the university, like those other kinds of mentorships or relatability that have been important for me, to have other Indigenous academics who are more senior so to speak.

MR: In reading these student essays over the years, I know Dawn and I have definitely noticed a lot of recent works in Indigenous studies, including your work, John. But there's also the opportunity to see older works being reread here. I was excited to see the University of Minnesota Press rerelease the Manuel and Posluns book *The Fourth World*. Can you speak to how older Indigenous works are being entered into large-scale academic conversations, after a period of them being not so well known?

GC: Yeah, I'd like to think that I helped get *The Fourth World* republished. It came out of a number of years of having to basically bootleg copies of it, or selections from that text, into the class, because of its remaining relevance. At the NAISA conference held at UBC, Jason Weidemann, who's my publisher, said they're starting to think of going through the backlog of not-available Native studies texts, and do you have any suggestions? And I said, immediately, the Posluns and Manuel book.

JB: When I was teaching at York University, Michael Posluns was my grad student doing his master's in environmental studies, and he spoke about his time writing that book, and he wanted to explore how Jewish philosophy could inform what he learned. Martin Buber, in particular, was important to his work. He was trying to bring out Buber's insights and reflect back on his work with George in those earlier days.

And so yes, there is contemporary relevance to the earlier literature. We can revisit that literature in light of more recent contributions. For example, in the mid-1990s, I was fortunate to supervise Harold Cardinal as my PhD student at UBC. He wrote *The Unjust Society* in 1969, and there's so much he wrote at that time that has relevance now. Harold did not shy away from putting his past ideas in a contemporary light. I like to watch students go back to that work and discover that it still speaks to them.

I've also heard that William Wuttunee's *Ruffled Feathers* is going to be republished. Wanda Wuttunee, his daughter, teaches at the University of Manitoba, and she is getting essays together to publish alongside it. You know, that book was controversial because it somewhat aligned Trudeau's and Chrétien's agenda, which was to get rid of the *Indian Act*. It's important to see why people like William Wuttunee might have seen this as a positive challenge, and it is important for students to grapple with the continued problematic role of the *Indian Act* in the present moment.

MR: For sure. I think when people do want to look at these ideas – like George Manuel and other writers from that era of the 1970s, the Red Power movements in the United States, Canada, and elsewhere – it is important. We're having to work against the obsession with the new. There's this progressivist narrative that intellectual thought is always moving forward, and therefore the emphasis is always on what are the most recent publications, and that's something that we have to work against.

One writer that comes up for me regularly is the late Mohawk scholar Patricia Monture-Angus. She's somebody who's been forgotten about. Of course, her work is very high profile, but at the same time, in a lot of the discussions now with respect to gender and the law, I feel sometimes that her work does kind of fall off the radar because, oh, it's from the 1990s? So now that we're thinking about Indigenous intellects from the 1970s, even from a couple of decades ago, yeah there's the sense, there's the confusion of looking at things that are old, or, you know, looking at intellectual work that is quote, unquote, "old" [*laughter*].

MR: My next question is built around Leanne Simpson's essay "Land as Pedagogy." We have a university that's progressively turned to online systems and methods, alongside the deepening of performance metrics and publication outputs. Against these tendencies, Indigenous forms of land-based pedagogy hold out an alternative, more grounded, model of university education. What are your experiences with land-based pedagogies working, or not working, with the academy?

GC: Well, my concern was originally that the move to online platforms would diminish land-based learning, which is really borne of a practical ethics that involves the repetition that Leanne teaches, and face-to-face learning is everything, the context of the land is everything. I've been lucky: over the last year we were able to secure some money for programming face-to-face, kind of a hybrid model.

We got a lot of the academic content and ethical considerations completed online and really focused on the land-based component of the knowledge production when we gathered on the land. We were also able to secure money to support programming, to keep people safe and out of harm's way. But there was always the push to have this type of education become at least a little bit more hybridized online, and I still am concerned with that. The funders and so on think that it's cheaper to be able to deliver this type of education on an at least hybridized online platform, and we've had to push back quite significantly against that tendency.

JB: You can't replace the land-based experience with the online. My sense of this is that, as a human, I'm to facilitate these experiences, but it's often the geese that are doing the teaching, or the butterflies, or the rock surface of the escarpment, or the currents in the water – this is where you look to take your cues and understand who we are, who they are, who we are together; it's not something that can just be imparted through a voice or a concept, there are other beings that have to be involved. You can't call geese into a Zoom session, right? You can't have the embodied, embedded, immersive experience just through a digital means. And that's not just speaking romantically, that's also speaking critically, you know. Things happen that are challenging in learning from earth, rock, water, animal, etc. So, I think there's a place for digital learning to help get people to certain points, but there's no replacement.

GC: There are also barriers to this type of medium to certain communities over others. Half of the communities that we serve are fly-in and don't have access to the Internet on a house-by-house basis. It was important to figure out ways to get our resources to those communities so that they could have their own time on the land. There was a lot of creativity that had to go into thinking about how to use the resources that we've been able to acquire from the state in order to redistribute back into the economy of knowledge production. We were able to do that, but it took thinking creatively about the mandate that we have as an institution and so on.

MF: It's a pretty big part of the experience of being urban and Indigenous. For people like myself who live in a city that is far away, there's always been that struggle of having to maintain connection and relationships remotely. I can go back and visit fairly regularly, but that's not a given, especially considering the high rates of poverty among urban Indigenous people.

I'm concerned about the rhetoric that when the pandemic is over, we want to return to normal. I don't want to assume that the existing pedagogies in universities were working for people before the pandemic. Within university there are inherently and structurally colonial capitalist hetero-patriarchal dynamics that have always been at play. As an Indigenous student, and as a neurodivergent person, the classroom environment has never worked for me, ever. I've always had to find ways to maintain connections with territory and with my family, often in ways that involve doing it remotely.

The university for me has always been an incredibly unempathetic place. And if anything, I think there are some positives to come out of this collective experience of the pandemic. People are appreciating that you can't always be expected to be a workaholic and just get it done regardless of what's going on outside you. It seems to be an assumption that your life and the world around you are not going to impact your ability to be a so-called productive student in our neoliberal dynamics. Now that there is this collective experience, I hope that there's space for a larger role of empathy within the university.

Up until this experience, accommodations were literally a pathologized thing. Now, having to seek out accommodations and drastically restructure how a class is done is the norm instead of the exception. People who feel especially at risk for COVID are having to push back – you know, saying, "No, online work doesn't work for me either, but I'm willing to make these adjustments if it means keeping safe."

I do think there is a potential for a more empathetic culture than there was before; it's one of the more bizarre positives that has come out of this. I just saw this recently with everything happening right now with the discovery of the different mass graves at former residential school sites. I've talked to a few family members and other Indigenous students who have had student services reach out to them and offer them financial support. This wasn't something that people had to ask for, and people weren't even asked to say how this experience of trauma directly impacted them financially – no questions asked, just offering support. I think these are the kinds of things that should be normal, and normalized.

SS: All of you have remarked on how the whole system is so heavily burdened with colonization. I'm wondering if the pandemic might have actually *helped* with some unlearning of ways. For a lot of small groups like RAVEN, we do focused work around decolonization and unlearning. But how do you build that into a system that's as big as a university?

MF: For a university to feel like a safe place, and a supportive place for Indigenous people to be who they are, think how they think, and be able to do research and write and speak on those issues in a way that's supported, all of this definitely requires a lot of change. And it's not something that's going to be solved by, for example, each individual student having to write it out and seeking accommodation every time something happens in their life that makes it difficult or next to impossible for them to finish a particular paper, let alone a whole semester.

GC: One thing universities haven't got right about supporting Indigenous scholars, especially younger ones doing community-based work, is their focusing so heavily on standard peer review, focusing on those traditional top-tier journals, single-author pieces, as opposed to more collaborative approaches.

JB: That's right.

SS: Well, this anthology features ten essays from over a decade of writing, and very different kinds of writing. In considering them all together for the first time, and rereading them, and just looking at the general arc of what the papers cover, because there are a number of topics on Indigenous

themes, what sorts of sustained engagements, or what surprises or open
questions, came out for you as you reconsidered this body of work? What
stands out for you?

JB: One thing that stands out in this work is how the students' essays talk to
one another from different disciplinary perspectives. Sometimes there are
silos in academia that separate us from understanding important themes
raised by other people. Yet here, in this book, the disciplines are cross-
cutting, in ways that have relevance for geography, and social work, and
law, and political science.

GC: Yeah, I want to second that. It was multidisciplinary, but the glue that
kept it as a cogent broader piece was Indigenous studies. That's pretty
cool to see. As a kind of a fighting creed that wasn't always accepted, still
isn't entirely accepted within the academy, it's cool to have it as a unifying
force, Indigenous thought and action.

MR: I spoke with one of the last RAVEN winners after telling him the news.
It was remarkable to recall how this winner was originally not even
sure he should send his work. He wasn't sure if it would meet the call
for "environmental issues," you know? He wondered if he'd qualify as
a mature student. He was not located at a big university. And then he
shared this simply beautiful story about naming in Anishinaabe culture
and growing up in a part of the world where he was taught to appreciate
the relations around him. My point is just that it seemed there was a lot
of discouragement, he felt a lot of discouragement in his part of the
academy.

MF: That's amazing that he did send it in, and those are exactly the kinds of
things that to me are inspiring. And I think there's hope, and again not
because I think things inherently get better, or move forward so to speak,
but it is a space where there are certainly a lot more Indigenous people at
universities than in the past. And not solely as undergraduate students, but
graduate studies, staff members in these different programs.

This is part of what I was thinking about when I was going over the
papers last night: yes this work is incredible, and I'm glad that it can be
encouraged through RAVEN, through the mentorship of faculty members
and instructors that people meet, and to get positive affirmation. And
that's related, unfortunately, to the work that still must be done in the
university, that this student felt uncomfortable or unsure if they should
even share the paper. It makes me sad and angry at the same time.

SS: Thank you for these generous reflections, all of you. Following on Mike,
I guess the final question from me is what would you like to see happen
with an anthology like this?

JB: I hope this book encourages – to use James Tully's word – "reciprocal
elucidation." This is what I find in this collection. We're not merely
isolated, or fragmented (though that can happen sometimes), we're also

a community. I hope people would recognize that being a community doesn't mean we're all the same, right? There's room for differences of opinion in communities. We need critique; there's room for different approaches to our work. I hope readers receive the book in this way.

GC: I just think about if it were somebody like me who got to see their work, or the work of one's peers, in university at that level, how empowering that would have been, and how much worth was put on one's words and one's labour, that it would be cool to have students see themselves in that work, in the classroom.

MF: What I hope people will take away from this is another example of the ways that we're able to articulate our ideas, our thoughts, our experiences on our own terms, as we see them, and in a way that's legible for ourselves and within our communities. And that doesn't feel the need to basically give in to a particular assimilationist dynamic, or to re-articulate what we say so that it's more palatable and more understandable, along terms of fundamentally non-Indigenous institutions. So hopefully that's what the collection will inspire: that people can move away from that and just express things on their own terms.

Raven Goes to School – (Re)learning Transformation from Graduate Students

SARAH HUNT – TŁALIŁILA'OGWA

Káwaṯilikálla saw a Raven strutting along the beach. All he could think of was how to get tribesmen. He said to the Raven: "I wish you were a man so that you could come and be my tribesman." The Raven threw back his feather dress and replied: "What am I, after all? You see I am a man when I wish to be." Said Káwaṯilikálla, "Well, you had better put that feather coat away. Come and be a man, and use this feather dress only when we dance. That is what I have done." This the Raven agreed to do.[1]

Turning to the left as I walk across the threshold of my office at UVic, I feel the feather dress resting gently upon my shoulders. Looking around, I orient myself here in the territories and homelands of the Lekwungen Nation. I have entered into the village of Sungayka, which I understand to mean "snow patches," a village of the Checkonien family.[2] In this sense, I have entered into a place of cultivation, trade, cultural and spiritual knowledge, in which the Checkonien family is my host. Before taking the floor here, I must consider how I am upholding the protocols and practices that recognize the governance of this family and the broader Lekwungen Nation. I am situated in an ancient and persistent Lekwungen ecology in which the natural world – the *tunulth*, or land beneath this building – the plants, birds, and waters around, above, and beneath me, are alive with the sacred wisdom. I raise my hands and recognize the Indigenous space we're in, enveloped by the feather coat upon my shoulders. Raven dances among the snow patches.

This is not just any raven, but the Raven ancestors that are part of the origin stories of the Kwakwa̱ka'wakw tribes from which I descend. Raven stories abound in the cultural practices of the Kwagu'ł, from which I descend through my grandfather. But the raven referred to in the origin story above is from the Dzawada̱'enux̱w, my grandmother's people. This is Raven back at the start of time, when our tribes were being formed on the beach – not just any beach, but the beach up the coast from here at the mouth of the Gwa'yi River. In this story, we see Kwakwaka'wakw epistemology as expressive of transformation, through the embodiment of spiritual roles that allow us to move between worlds and take different forms depending on the space we're in. Understanding the university to be a Lekwungen space, an Indigenous space, then, I ruffle my feathers and circle around to the left as we do when we enter into the big house to do sacred governance work. Embodying an extensive network of relations across coastal nations, I orient myself towards the Lekwungen families and elders to whom I am accountable. It is these teachings that direct me in my work here, and that also guide many of the graduate students on this campus who are engaged in nurturing relationships with Indigenous peoples, knowledges, and lands.

As Songhees Elder Skip Dick reminds us, we need to ensure we don't take shortcuts in protocol in order to keep our ancestors on the other side happy.[3] Working in the university as an Indigenous place, then, can be a continuation of ancestral relations that have existed among our people since time immemorial. Ravens from many corners of the globe show up, share food, awkwardly talk in different languages, trying to make sense across their teachings, telling tales of their journeys across diverse lands and worlds. We squawk, we preen, we go down to the shoreline to cleanse ourselves. Our shiny feather coats gleam in the sunlight.

But quickly, it becomes clear that we are not only in an Indigenous space. At the university, we are also in a colonial space that continues to be contoured and disciplined by structures of power, authority, value, and measurement, and that constrain the full expression of Indigenous ways of being. Looking out my window, I see a native plant garden that is tended to by non-Indigenous university staff. There are no markers, no practices to connect these culturally significant plants to the governance systems in which they have flourished for thousands of years – no Lekwungen people harvesting medicines, no Indigenous plant names being used to encourage their growth across the seasons. Where are the women and elders, the heads of the Checkonien family, I wonder? I open my email and am bombarded with messages discussing diversity, equity, and inclusion surveys, strategic plans on Indigenization, requests to sit on sustainability committees, invitations to review papers on environmental issues inspired by Indigenous peoples without actually referencing Indigenous scholars – Raven is no longer dancing.

Where are we now? It seems we have moved into another space, a subtle shift across realms.

Turning left again to signify leaving the sacred space of shared governance where our connections to land are affirmed, I remove the feather dress and put it away in order to protect the power of those Raven ancestors. Raven is not at school anymore. I have transformed as I shift into the space of the colonial institution in which my recognition as an Indigenous faculty member is shaped by a history of marginalization, racism, sexism, and misrepresentation. But working in *this* space is also sacred, in a sense, because it requires supporting the activism and demands for structural change that students have long pushed for. Student activism continues to drive much of the necessary radical and transformative work at universities and in our disciplines and schools, as realities of settler colonialism, racism, and hetero-sexism push up against the limitations of the slow pace of institutional change.

The student essays in this volume speak, in direct and indirect ways, to these tensions inherent to centring Indigenous peoples, knowledges, lands, and relationships while navigating institutional norms predicated on intersecting dynamics of colonialism. The voices of Indigenous students and students of colour, in particular, point towards the creative ways students continue to insist on making their voices heard, on their own terms, despite being in programs without Indigenous and racialized women faculty,[4] or not being taught that Indigenous genealogies and oral histories, including our own, do indeed count as "ecological" or "environmental" enough.[5]

These essays allow us to witness the new forms of representation students are bringing into our disciplines and institutions, insisting on avoiding the reproduction of environmental language that further normalizes colonial perspectives (e.g., Atlanta Grant's use of "food cycling," which centres Indigenous philosophies of relationality).[6] We are also witness to the ways Indigenous storytelling, theorizing, and cultural norms are being danced alive here; we see Raven dancing across the page, along with the relations who make up entire cosmologies that are the foundation of our nationhood. Yet it is important to continue asking about the costs students bear in choosing to dance their laws alive – or the laws of the nations with whom they work – within the academy.

As a queer Indigenous woman whose scholarship has largely focused on the gendered nature of justice, I am acutely aware of the way institutional and disciplinary norms continue to perpetuate colonial hetero-patriarchy. This impacts students in both subtle and overt ways, shaping their movement through graduate programs and beyond. As a professor, graduate supervisor, committee member, and teacher I am accountable to the students whose journeys across institutional-community spaces I have witnessed, and I draw on the bold and transformative work of some of those students here. Indigenous graduate students, particularly women and gender-diverse students, continue to talk, write, and create art and theory that reflects the harm they endure even as they centre their loving relations within academia. As Indigenous women and gender-diverse relatives enter into colonial institutions, embodying their ancestral knowledge and community stories, they describe the transformation of their love and labour through settler-colonial frames of recognition, turning revolutionary work into something "palatable, acceptable, and exploitable."[7] Indeed, universities are particularly at risk of cherry-picking only the aspects of Indigenous knowledge they deem desirable in this era of reconciliation – with an expectation of conciliatory or pacifying approaches.[8]

As John Borrows discussed in his introductory chapter, universities are making significant changes in the treatment of Indigenous knowledge and people through consideration of the Truth and Reconciliation Commission's Calls to Action. Yet we must remain critical of the limitations of these tools, asking how well they address the everyday realities of gender-diverse and sexually-diverse Indigenous and racialized students; this urgent and persistent tension is, I feel, at the very centre of our structural entanglements with settler colonialism. The Calls to Action restrict discussion of gender to Call to Action 41, which calls for a government enquiry and investigation into gendered violence.[9] Now that that enquiry is over, have the gendered impacts of residential schools been resolved? What might a gendered lens offer to efforts to bring "reconciliation" into Canadian universities, particularly where Indigenous studies and ecological justice are concerned? The UN Declaration on the Rights of Indigenous Peoples, another mechanism being taken up by post-secondary institutions, particularly in light of the creation of the *Declaration on the Rights of Indigenous Peoples Act* in British Columbia, has also been criticized for only mentioning women in three of forty-six articles and for taking a deficit approach.[10] These gaps are significant not only because of the gendered nature of dispossession, but also because of the gendered nature of Indigenous legal orders that underpin our nationhood. Our graduate programs must seek out ways to work within and across these gaps, seizing the opportunity to ask new and different questions. How can we transform our institutions, our disciplines, and our work with Indigenous nations such that gender and sexual diversity is normative? What happens when we put this question at the centre of our considerations of reconciliation and decolonization?

Madeline Whetung proposes that in this age in which Indigenous people are being asked to reconcile, we engage instead in "unreconciliation" – a political act premised both on the recognition of colonial violence and the recognition that there is limited possibility for repair.[11]

This aligns with Rachel Flowers's observation that Indigenous women's resistance is often conflated with love, thereby avoiding the necessity of confronting women's refusal to forgive in the context of ongoing hetero-patriarchal violence across settler-colonial spaces – including universities.[12] More than just making space for uncomfortable feelings, Flowers, Quill-Peters, and Whetung provide crucial insights into the transformative potential of the felt theories of graduate students who make astute interventions into the structural constraints placed on Indigenous peoples and knowledge in the era of reconciliation.[13]

For graduate students who are experiencing hetero-sexism, transphobia, or racism within their everyday movement through our universities, scholarly associations, conferences, and sites of publication, we must take seriously students' caution against an expectation of repair, reconciliation, or good feelings. Instead, Whetung sees "remaining unreconciled" as a way of insisting on the examination of the harmful or violent relations of power we are enmeshed in.[14] For those of us working in universities, we need only look to the fact that these institutions sit on lands stolen from their host nations to understand the complicated truths brought to the fore by remaining unreconciled. As UVic faculty, this requires continuing to look for ways to act in responsibility to the Lekwungen women who governed cultivation here – gendered forms of governance that are specific to this place. Aligning my thinking and action with Whetung, then, remaining unreconciled is a way "to hold space to imagine a different type of relationship from where we are."[15] Colleagues at all levels of our universities would do well to practise the self-reflexivity that facilitates the asking of hard and uncomfortable questions, as modelled by these graduate students, looking for opportunities to take on some of the affective labour required to unsettle the stubborn cultural norms of the university.

As I take my feather dress off the hook once more, I reflect on how the brave actions of these graduate students have made it possible to dance our laws alive in this space that is at once Sungayka and the University of Victoria. Raven dances across the snow patches, walks into the classroom, and greets diverse relatives from near and far. Here, in this space between worlds, all are welcome.

NOTES

1 Edward S. Curtis, *The North American Indian*, vol. 10. *The Kwakiutl* (Norwood, MA: Plimpton Press, 1915), 132–4.

2 Cheryl Bryce in *University of Victoria Campus Plan* (Victoria, BC: UVic, 2016), 10, https://www.uvic.ca/campusplanning/assets/docs/CampusPlan2016 .pdf.

3 This was shared in personal and public conversation numerous times by Elder Skip Dick, Songhees Nation (Lekwungen).

4 See, for example, Erica Isomura's essay in this volume, "Swimming Upstream against (Neo)colonialism: On Salmon Aquaculture Supremacy and the Decline of Sockeye in the Stólō."

5 See Wade Houle in this volume, "My Story."

6 See Atlanta Grant in this volume, "The Berry Picker."

7 Quill Violet Christie-Peters, "Hands Off! On Indigenous Women's Love and Labour in the Institution," in *Every. Now. Then. Reframing Nationhood*, exhibition catalogue (Toronto: Art Gallery of Ontario, 2017),

8 Michelle Daigle, "The Spectacle of Reconciliation: On (the) Unsettling Responsibilities to Indigenous Peoples in the Academy," *Society and Space* 37, no. 4 (2019): 703–21. For a deeper understanding of navigating reconciliatory discourse and expectations as an Indigenous woman and descendent of residential school survivors, see Michelle Daigle's discussion of examples spanning across her time as a doctoral student, postdoc, and professor.

9 *Truth and Reconciliation Commission of Canada: Calls to Action* (Winnipeg: Truth and Reconciliation Commission of Canada, 2015), https://www2.gov .bc.ca/assets/gov/british-columbians-our-governments/indigenous-people /aboriginal-peoples-documents/calls_to_action_english2.pdf.

10 Brenda L. Gunn, "Bringing a Gendered Lens to Implementing the UN Declaration on the Rights of Indigenous Peoples," in *UNDRIP Implementation: More Reflections on the Braiding of International, Domestic, and Indigenous Laws* (Waterloo, ON: Centre for International Governance Innovation, 2012), 33–8.

11 Madeline Whetung, "On Remaining Un-reconciled: Living Together Where We Are" (unpublished working paper, University of British Columbia, 2017).

12 Rachel Flowers, "Refusal to Forgive: Indigenous Women's Love and Rage," *Decolonization: Indigeneity, Education and Society* 4, no. 2 (2015): 32–49.

13 Diane Million, "Felt Theory: An Indigenous Feminist Approach to Affect and History," *Wicazo Sa Review* 24, no. 2 (2009): 53–76.

14 Whetung, "On Remaining Un-reconciled."

15 Whetung, 18.

Contributors

Helena Arbuckle wrote her chapter from the perspective of a woman with Celtic (Scottish and Irish) and English ancestry, and as a settler who presently calls Ləkʷəŋən, Hupačasath, and c̓išaaʔatḥ lands home. She recognizes her privileged experiences and opportunities at the expense of Indigenous lands, lives, world views, and livelihoods, and encourages fellow settlers to learn the long-standing story of genocide and disposses-

sion that brings us to reside on forcibly taken lands.

André Bessette nikshihkaahshoon. Niwah komahkanak apihtokohshan, mistogoso, irish, ekwa croatian. I'm an urban Métis person from a family that has been disconnected from our Métis culture because of coloniza-tion. Since I was a young adult, I have been working with my family to reclaim our histories and cultures. I have been a visitor on the traditional and unceded territories of the Musqueam, Squamish, and Tseil-Waututh Nations for most of my life. I practise solidarity with our host nations by connecting with community members and organizations, informing other visitors of the local impacts of colonization, and acting alongside our communities in political action. As a Métis individual and citizen of Métis Nation BC, I reject all claims made by Métis organizations on the lands and resources of sovereign Indigenous nations without consent from those nations. I look forward to being part of the work to build relationships of solidarity and respect between Métis and First Nation communities. I thrive when I'm participating in social justice, Métis jig-ging, politics, policy, and being an overall nerd.

John Borrows BA, MA, JD, LLM (Toronto), PhD (Osgoode Hall Law School), LLD (Hons., Dalhousie, York, SFU, Queen's & Law Soci-ety of Ontario), DHL, (Toronto), FRSC, OC, is the Loveland Chair in

Indigenous Law at University of Toronto Law School. His publications include *Recovering Canada: The Resurgence of Indigenous Law* (Donald Smiley Award for best book in Canadian political science, 2002), *Canada's Indigenous Constitution* (Canadian Law and Society Best Book Award, 2011), *Drawing Out Law: A Spirit's Guide* (2010), *Freedom and Indigenous Constitutionalism* (Donald Smiley Award for best book in Canadian political science, 2016), *The Right Relationship* (with Michael Coyle, ed.), *Resurgence and Reconciliation* (with Michael Asch, Jim Tully, eds.), and *Law's Indigenous Ethics* (2020 Best Subsequent Book Award from Native American and Indigenous Studies Association, 2020 W. Wes Pue Best Book Award from the Canadian Law and Society Association). He is the 2017 Killam Prize winner in Social Sciences and the 2019 Molson Prize winner from the Canada Council for the Arts, recipient of the 2020 Governor General's Innovation Award, and the 2021 Canadian Bar Association President's Award winner. He was appointed as an officer of the Order of Canada in 2020. John is a member of the Chippewa of the Nawash First Nation in Ontario, Canada.

Da Chen currently lives in Toronto, the traditional territory of many nations, including the Mississaugas of the Credit, the Anishinaabeg, the Chippewa, the Haudenosaunee, and the Wendat peoples. He completed his MSc in planning at the University of Toronto in 2020 with a focus on Indigenous and environmental planning. He also holds a BA (hons.) in city studies and political science. He is a first-generation immigrant of Chinese descent, which shaped a lot of his upbringing and helped inform a lot of his understanding of the world and his relationships with it. He enjoys spending time outdoors, reading books and writing poetry, and learning more about his role in this world.

Glen Coulthard is an associate professor in First Nations and Indigenous Studies and in the Department of Political Science at the University of British Columbia. Glen has written and published numerous articles and chapters in the areas of contemporary political theory, Indigenous thought and politics, and radical social and political thought (marxism, anarchism, post-colonialism).

Heather Dorries is an assistant professor jointly appointed to the Department of Geography and Planning and the Centre for Indigenous Studies at the University of Toronto. Her research focuses on the relationship between urban planning and settler colonialism and examines how Indigenous intellectual traditions – including Indigenous environmental knowledge, legal orders, and cultural production – can serve as the foundation for justice-oriented approaches to planning. She is of Anishinaabe and settler ancestry and a member of Sagkeeng First Nation in Treaty 1.

Michael Fabris is an assistant professor of geography at the University of British Columbia. His research focuses on processes of dispossession in settler colonial contexts, and the Indigenous social movements that seek to challenge these processes by reasserting Indigenous land and water relationships through their own forms of governance and legal jurisdiction. He is of Blackfoot and mixed European descent, affiliated with the Piikani Nation.

Tosin Fatoyinbo is a lawyer with two master's degrees in public policy and human rights. Tosin, an immigrant in Canada, is curious about the original owners of the land on which he now lives. He is discovering that Indigenous legal systems, like other legal systems, can evolve and should be respected rather than subjugated.

Atlanta Grant (she/her) is a writer and systems change worker. She is an MA graduate from the University of British Columbia, where her research focused on utilizing decolonizing methodologies in cross-cultural research and practice. Her writing focuses on poetry as a form of storytelling and identity-making, interweaving research topics that surround cross-cultural collaborative intervention and Indigenous knowledge systems. Currently she works in Data and Research under Social Policy in Arts, Culture, and Community Services with the City of Vancouver, and speaks at various research conferences, encouraging researchers wishing to navigate cross-cultural spaces to begin the internal work necessary for safe and abundant community partnerships and project support.

Erica Hiroko Isomura is a writer and interdisciplinary artist. She was raised by a Cantonese Canadian mother and a sansei Japanese Canadian father on Qayqayt, Musqueam, Squamish, and Tsleil-Waututh territories beside the Stó:lō (New Westminster, BC). Her work has appeared in *Canthius*, ArtsEverywhere.ca, *The Fiddlehead*, and *Vallum*, among other publications. Erica holds a BA in social sciences from the University of Victoria and is currently an MFA candidate in creative writing at the University of Guelph.

Dawn Hoogeveen BA, (Carleton) MA, (Simon Fraser University) PhD (University of British Columbia), is a university research associate in the Faculty of Health Sciences at Simon Fraser University, cross-appointed with the First Nations Health Authority, where she works as a senior research fellow. Her publications range from academic journal articles in human geography and Indigenous studies on mining laws and settler colonialism, to climate change equity work and critical research on environmental assessment and gender, health, and wellness. She completed postdoctoral work in the Institute for Resources Environment and Sustainability (UBC, 2018) and in health sciences and the Geography Program at the University of Northern British Columbia with the Environment Community Health Observatory (2020). Dawn is a third-generation Canadian settler of Dutch and British ancestry, raised on Anishinaabe territory on Williams Treaty lands near Peterborough, Ontario, who is currently an uninvited guest on Coast Salish territory in Vancouver, BC.

Wade Houle is in his final year in the Faculty of Graduate Studies Master's Program at Brandon University. Wade resides and works on Treaty 2 territory in Manitoba, the traditional lands of the Anishinaabek, Cree, and Dakota peoples, and the homeland of the Métis Nation. He is Anishinaabe and Métis and has been an educator for fifteen years and is passionate about Indigenous education. Wade believes in an introspective approach to leadership where one is grounded in their experiences and that being anti-racist and anti-oppressive is key for the future.

Sarah Hunt/Tłaliłila'ogwa, BA, (University of Victoria), MA, (University of Victoria) PhD (Simon Fraser University), is Canada Research Chair in Indigenous Political Ecology and an associate professor at the University of Victoria School of Environmental Studies. Sarah has published upwards of forty journal articles, book chapters, and reports on a range of topics pertaining to questions of justice for Indigenous people. Her writing has been published in journals such as *Geography Compass, Atlantis, Professional Geographer,* and *Cultural Geographies,* and her recently authored or co-authored reports include *Access to Justice for Indigenous Adult Victims of Sexual Assault* (with Patricia Barkaskas), *An Introduction to the Health of Two-Spirit People,* and *Indigenous Communities and Family Violence: Changing the Conversation* (with Cindy Holmes). Her writing can be found in anthologies such as *Indigenous Research: Theories, Practices and Relationships; Keetsahnak: Our Missing and Murdered Indigenous Sisters; Determinants of Indigenous Peoples' Health in Canada;* and *The Winter We Danced: Voices from the Past, the Future, and the Idle No More Movement.* Sarah is a 2014 recipient of the Governor General's Gold Medal Award, 2017 recipient of the Glenda Laws Award for Social Justice from the American Association of Geographers, and 2022 recipient of the President's Distinguished

Alumni Award from the University of Victoria Alumni Association. Sarah/Tłaliłila'ogwa is Kwakwa̱ka'wakw – Kwagu'ł through her paternal grandfather Chief Henry Hunt, and Dzawada'enuxw through her grandmother Helen Nelson, and is also Ukrainian and English through her maternal grandparents.

Danette Jubinville is a mother, birth worker, and researcher based in xʷməθkʷəy̓əm, S̱ḵwx̱ wú7mesh, and səl̓ílwətaʔ territories (Vancouver, BC). Danette is currently a PhD candidate at the SFU Faculty of Health Sciences and the director of research and education for the Ekw'í7tl Indigenous Doula Collective. Danette's work aims to advance reproductive justice for Indigenous people in the city, a project near to her heart as an urban mixed-blood woman and mother with Cree, Saulteaux, French, Jewish, German, and Scottish descent.

Kevin Ly was born and raised in what is colonially known as Vancouver, Canada, on the traditional, ancestral, and unceded territory of the Coast Salish peoples – S̱ḵwx̱wú7mesh (Squamish), Stó:lō, and Səl̓ílwətaʔ/Selilwitulh (Tsleil-Waututh) and xʷməθkʷəy̓əm (Musqueam) Nations. He was privileged to attend the University of British Columbia, where he studied international relations and geography, with a focus on environment and sustainability, the latter of which provided him with the foundational knowledge to pursue the topic of his essay. As the child of refugees who escaped the Vietnam War, Kevin was compelled to write his essay after researching the contrast in the treatment of and policies towards immigrants/war refugees and that of Indigenous peoples, and the intersection with natural elements, such as water.

Laura Peterson is from the K'ómoks First Nation on eastern Vancouver Island, BC, who also has Tahltan and Métis ancestry. She grew up in the Comox Valley and Victoria area, and later obtained a BA from Eastern Washington University and an MA in Native studies from the University of Alberta. Laura's MA research explored a traditional overland trail in Wood Buffalo National Park in northern Alberta. She chose this topic due to the poor condition of these trails and to preserve knowledge associated with their use and importance. Working with local Cree and Chipewyan knowledge holders and youth, her project supported the transition of knowledge and skills between generations. Laura joined the Parks Canada team in 1998 as the cultural resource management adviser working in Fort Smith, NT. She met her husband, George, and they raised two children, Niklas (seventeen) and Amber (twenty-five). Now back on Vancouver Island, she works at Gulf Islands National Park

Reserve in Sidney, BC. Laura enjoys spending time on the land. Her interests include canoeing and hiking with family and friends, playing hockey, and travelling.

Max Ritts BA, (McGill), MA, (University of Toronto), PhD (University of British Columbia), is an assistant professor in the Graduate School of Geography, Clark University. He has authored over a dozen scholarly works and is completing a monograph about sound, industrial development, and the politics of nature in Ts'msyen Territory, entitled *A Resonant Ecology* (Duke University Press, 2024). Since 2013, Max has collaborated with the Gitga'at First Nation across several capacities, including as a community newsletter editor, research technician, and a youth coordinator. Max is of mixed Jewish descent and was born and raised in Toronto (traditional territory of the Mississaugas of the Credit, the Anishinaabeg, the Chippewa, the Haudenosaunee, and the Wendat peoples).

Susan Smitten BJ honours (Carleton) retired from her role as the executive director of RAVEN (Respecting Aboriginal Values and Environmental Needs). She built RAVEN from its origins in 2009, and it is the only non-profit charitable organization in Canada with a mandate to raise legal defence funds to assist Indigenous peoples who choose to enforce their rights in court to protect their traditional lands, waters, and way of life. Susan is also an award-winning filmmaker and writer whose past projects communicate the connection between environmental issues and First Nations' stewardship of the land. For RAVEN, Susan directed the acclaimed *Blue Gold: The Tsilhqot'in Fight for Teztan Biny*, giving voice to the Tsilhqot'in people's unanimous rejection of Taseko Mines Limited's "Prosperity" open pit gold-copper mining project. She co-directed *Wild Horses, Unconquered People*, about the Xeni Gwet'in First Nation's monumental rights and title legal challenge. She is also the author of several books of ghost lore (Lone Pine Publishing). Susan is a second-generation settler of mixed European ancestry, born and raised in Tkaronto, in the Dish with One Spoon Treaty territory. She currently lives with her husband and daughter in the traditional territories of the ləkʷəŋən peoples.

Index